Astronautical Engineering

Series Editor
Scott Madry

More information about this series at http://www.springer.com/series/5495

New Space Ventures

Joseph N. Pelton

Space 2.0

Revolutionary Advances in the Space Industry

Springer

Published in association with
Praxis Publishing
Chichester, UK

PRAXIS

Joseph N. Pelton
International Association for the Advancement of Space Safety
Arlington, VA, USA

ISSN 2365-9599 ISSN 2365-9602 (electronic)
Springer Praxis Books
ISBN 978-3-030-15280-2 ISBN 978-3-030-15281-9 (eBook)
https://doi.org/10.1007/978-3-030-15281-9

This Springer imprint is published by the registered company Springer Nature Switzerland AG.
The registered company address is: Gewerbestrasse 11, 6330 Cham, Switzerland

Acknowledgments

A book that covers a wide range of space technologies and services clearly benefits from consultations with a broad number of experts and advisors. In the past year I have consulted with many friends and colleagues on space technologies, systems, services, and policy and would like to thank them for their valuable advice and counsel. Thus I would like to express my sincere appreciation to Dr. Scott Madry of the University of North Carolina, Dr. Ram Jakhu of McGill University, Dr. Margaret Whitehead of the Hudson Institute, Dr. Henry Hertzfeld of George Washington University, Michael Miniero of the Outer Space Subcommittee, U.S. Congress, Dr. Xavier Pasco, Director of the Strategic Research Foundation of France, Dr. Jean-Jacques Tortora of the European Space Policy Institute, Tommaso Sgobba of the International Association for the Advancement of Space Safety, Dr. James Green of NASA, Dr. Angelina Buckley of the Aerospace Corporation, Dr. Michael Kezirian of the International Space Safety Foundation, Dr. Su-yin Tan of the International Space University, Dr. Peter Martinez of the World Security Foundation, and many others who have also been helpful that I have unintentionally overlooked.

I would like to thank the dynamic duo of Maury Solomon and Hannah Kaufman of Springer that are, as always, a joy to work with and extremely helpful with the editing and production of my books.

Finally I would like to give a very special thank you to my friend and colleague Peter Marshall. Peter has served in many roles as a frequent coauthor and also as top quality editor. Once again he has helped to make this book, and I believe at least ten others, much more cogent and well parsed. Thank you, Peter.

Contents

About the Author

Joseph N. Pelton, Ph.D. is the former Dean and Chairman of the Board of Trustees of the International Space University. He also is the Founder of the Arthur C. Clarke Foundation and the founding President of the Society of Satellite Professionals International—now known as the Space and Satellite Professionals International (SSPI). Dr. Pelton currently serves on the Executive Board of the International Association for the Advancement of Space Safety. He is the Director Emeritus of the Space and Advanced Communications Research Institute (SACRI) at George Washington University, where he also served as Director of the Accelerated Master's Program in Telecommunications and Computers from 1998 to 2004. Previously he headed the Interdisciplinary Telecommunications Program at the University of Colorado-Boulder. Dr. Pelton has also served as President of the International Space Safety Foundation and President of the Global Legal Information Network (GLIN). Earlier in his career he held a number of executive and management positions at COMSAT and INTELSAT, the global satellite organization where he was Director of Strategic Policy. Dr. Pelton has been speaker on national media in the United States (PBS NewsHour, Public Radio's All Things Considered, ABC, and CBS) and internationally on BBC, CBC, and FR-3. He has spoken before Congress, the United Nations, and delivered talks in over 40 countries around the world. His honors include the Sir Arthur Clarke International Achievement Award of the British Interplanetary Society, the Arthur C. Clarke Foundation Award, the ICA Educator's award, the ISCE Excellence in

Education Award, and being elected to the International Academy of Astronautics. Most recently, in 2017, he won the Da Vinci Award of the International Association for the Advancement of Space Safety and the Guardian Award of the Lifeboat Foundation. Dr. Pelton is a member of the SSPI Hall of Fame, Fellow of the IAASS, and Associate Fellow of the AIAA. He is a widely published author with some 50 books written, coauthored or coedited. His Global Talk won the Eugene Emme Literature Award of the International Astronautics Association and was nominated for a Pulitzer Prize. His most recent books include *The New Gold Rush: The Riches of Space Beckon, Global Space Governance: An International Study,* and the second edition of *The Handbook of Satellite Applications,* all published by Springer. As Director of Project Share, while heading Strategic Policy for Intelsat, he played, a key role in the launching of the Chinese National TV University that now is the world's largest tele-education program. He received his degrees from the University of Tulsa, New York University, and from Georgetown University, where he received his doctorate.

1 The World of Space in Flux

Introduction

The space industry is in a period of enormous change. Newly emerging space technology and a myriad of innovative space applications are both in a state of unparalleled growth and market flux. Only the fields of computer science and artificial intelligence (AI) – the so-called cyber-industries – have a similar pattern of rapid technological advancement and industrial ferment. This is not at all surprising in that the space industry has largely become a software-defined extension of the computer industry. Communications satellites, remote sensing and weather satellites, and space navigation satellites currently represent the largest space applications industries. If one stops to analyze the underlying technologies on which these space applications are based, one finds that they are essentially digital processing and sensing space systems in the skies, using specialized digital software and hardware. Digital processing is at their core.

A further look reveals that 'disruptive' space technology, services and applications are being created by what might be called the 'Silicon Valley' effect. Bright young people out in Silicon Valley, and in other parts of the globe where the cyber revolution has spread, all have a new mentality. They are saying: "Let me re-invent the world." And, indeed they are doing just that.

This time of rapid change in the commercial space industry has been called "NewSpace" and "Space 2.0." Regardless of what one calls it, there are suddenly new ways of designing, manufacturing, testing and launching satellites. There has been a radical change in the markets and applications that space systems serve, and the why, where, when and how of the space industry has changed. New space ventures are being started on the basis of crowd-sourcing and 'Kickstarter,' an Internet funding platform for creative endeavors. Many Space 2.0 ventures are launched via financing obtained through the Internet.

New organizations with names such as SpaceX, Blue Origin, Virgin Galactic,

J. N. Pelton, *Space 2.0*, Astronautical Engineering,
https://doi.org/10.1007/978-3-030-15281-9_1

Kymeta, One Web and Planet are reinventing the commercial space industries. As a result, well-established space industries such as Lockheed Martin, Boeing, Airbus, Northrop Grumman and McDonald-Detwiler are reinventing themselves in order to keep up with the commercial space revolution.

A recent study found that more than 80 angel and venture-backed space companies have been founded since 2000 in this realm of Space 2.0. Indeed, several of these new firms are today big bucks operations and on the verge of becoming billion-dollar concerns. In this book we aim to examine what has allowed these NewSpace ventures to get off the ground – figuratively and literally. In this study we came up with four key factors. These critical success factors were listed as: "(1) Business Philosophy – creating and living an entrepreneurial spirit; (2) Financing – access to early stage risk capital and venture funding; (3) Technology Management – focus on 'spinning-in' technologies and Information and Communication Technology processes; and (4) Framework Conditions – favorable political and legal conditions supporting commercialization" [1].

And some are looking well beyond near-term growth of a new space economy. Stephen Hawking, Elon Musk and even Jeff Bezos have looked into the future and to centuries beyond. Elon Musk has envisioned a million people on Mars. An Asian entrepreneur is planning to build condos on the Moon in future decades. Jeff Bezos has looked to the future in a very logical and pragmatic way: "The Earth is finite, and if the world economy and population is to keep expanding space is the only way to go" [2].

Anyone with vision will see many parallels between outer space and the so-called New World of five centuries ago. But the real challenge is not to create a new space economy for humans in the centuries ahead. That much is fated. Space products and services will grow rapidly in the decades ahead. This is clear to anyone with a vision of the future. No, the challenge is not to repeat the mistakes of the past. Industrial and national leaders need to develop regulatory and legal systems and 'rules of the road' that allow us to not pollute outer space. We need to find a way forward to achieve the longer-term sustainability of outer space activities. In this regard the continuing buildup of space debris in Earth orbit, the ozone hole and pollution of the stratosphere, all suggest an inauspicious start. Without effective guidelines we could end up with space debris around the Moon and several of our nearest planets.

The space economy will, within a decade or so, grow from about $350 billion to a trillion dollar economy. Space products and services will mushroom as global demand grows and we exhaust many of Earth's resources. The real challenge is to undertake this expansion with some sense of ecological care and a wisdom that comes from learning from past mistakes. It is indeed a bit of historical irony that a part of the new space economy relates to recovery from global warming, coping with and even recycling space debris, and other environmental-recovery activities [3].

The Small Satellite Revolution

Students from Stanford University built a new remote-sensing satellite system of 'small sats' that they called 'Skybox.' This network was then renamed 'Terra

Bella' when these young entrepreneurs sold their network to ABC/Google for many millions of dollars in profits. In a tangentially related story, students who attended the International Space University who were inspired by Professor Scott Madry decided to create a new small satellite network that they named Planet Labs. Their three-unit cube satellite Earth-sensing satellites they call 'doves.' This is because when these small sats have their solar array systems deployed on each side in space these three-unit-cube-satellites – about the size of a pigeon – sort of look like birds.

From their mansion in Mountain View, California, these former students are creating a highly capable global satellite system. This growing network of small sats images the entire world in less than a day and generates a huge amount of data. Their network, that has now been expanded to include the satellites from Skybox/Terra Bella in a deal that links them to ABC/Google, has been renamed simply 'Planet.' This amazing start-up company managed to arrange for the launch of eighty-eight of their Dove satellites on a single launch via an Indian rocket on Valentine's Day, February 14, 2017. This 'flock' of 104 cube satellites, each weighing under 5 kilograms (about 10 lbs.), were launched over a span of 18 minutes. This unprecedented number of small sat launches set a new world record for the number of sa tellites launched by a single rocket launcher. The previous record was 39 satellites on a Russian launcher [4]. (See Fig. 1.1.)

Fig. 1.1 The record-setting launch of 104 small satellites on the Indian Polar Satellite launch vehicle on February 14, 2017. (Photo courtesy of the Indian Space Research Organization.)

Complex and Conflicting Trends in Satellite Communications

And these are just two stories about satellite networks developed for remote-sensing services. New developments in the communications satellite field are rampant. Here innovations are spewing forth in a wide range of different ways. At the high end of the commercial satellite market, some of the largest companies are developing huge new satellites that have ten to fifty times the capability of those that were launched just a few years ago. These satellites, known as high throughput satellites, have the ability to transmit at digital rates as high as 140 to 150 gigabits/second. They are being launched by companies such as Intelsat (i.e., their Epic satellites), Echostar/Hughes Network Systems (i.e., their Jupiter satellites), Via Satellites (i.e., Viasat 1 and Viasat 2), and Inmarsat (i.e., their Express satellites) [5].

This high end of the market that involves the deployment of geosynchronous satellites with masses of 5,000 kilograms upward are primarily to support television distribution and broadband data communications, typically using either Digital Video Broadcast (DVB) standards or Digital over Cable System Interface Standard (DOCSIS). These monster satellites have huge solar arrays that can generate well over 10 kilowatts of power and have large antennas that can be used by efficient feed systems to generate even hundreds of beams to allow frequency re-use – just as cellular mobile communication antenna systems do on the ground.

These geosynchronous satellites have become more and more efficient by using higher and higher frequencies, having more on-board power, finding ways to reuse frequencies, plus using multiple digital coding techniques to achieve these higher and higher digital throughput rates.

Thus, at one end of the spectrum there are now massive high throughput satellites (HTS) that require on the order of a quarter-billion dollars to build and launch. These satellites, which orbit almost a tenth of the way to the Moon, can transmit tens of thousands of television and high data rate channels to low cost very small aperture terminals on the ground and require no expensive tracking capabilities. The satellites are big, powerful and expensive, which allows millions of ground antennas – even down to hand-held units – to be very low cost.

Then, at the other end of the spectrum, we have new small satellite constellations. These satellites are designed to operate in low Earth orbit some 40 times closer to Earth than the big geosynchronous satellites. The concept is to deploy a lot of smaller satellites in large-scale constellations of hundreds if not thousands of spacecraft. Such a constellation can then blanket Earth and provide low latency (i.e., minimal delay) broadband data links. These small satellites are thus well-suited for Internet-type services. The small satellites in this case, however, are much bigger than simple cube sats (which are a very compact 10 cm × 10 cm × 10 cm in size – or about the size of a softball).

These types of smaller satellites for communications are thus about 150 to 250 kilograms in size. That makes them 30 to 50 times larger than a Planet 3-unit cube satellites. (See Fig. 1.2.) Yet they are nevertheless some 30 to 50 times

Fig. 1.2 A Dove 3-unit cube sat, about the size of a small bird. (Graphic courtesy of Planet.)

smaller than the giant high throughput satellites like an Intelsat Epic or Viasat 2 satellite. These small satellites for communications services need to be bigger to have higher power, to have bigger antennas to create spot beams needed for frequency reuse and also to have the capability to be pointed with accuracy.

The bottom line is that we are now seeing a great diversity of satellite design and new launch systems to support these quite different types of satellites that range from tiny so-called cube sats, to large-scale constellations of "small" satellites that are 30 to 50 times larger than cube satellites, up to giant high throughput communications satellites that are 30 to 50 times larger still than the small satellites in the low Earth orbit constellations.

Figs. 1.2, 1.3A, 1.3B, and 1.4 show the appearance of these tiny, small and huge satellites which are representative of the great diversity and radical changes that characterize the space industry today.

And the changes that are occurring in the field of satellite communications are more complex and complicated than just the size, mass, shape and capabilities of the satellites. There are a host of other innovations that are leading to key changes in the space industry.

Key Technical Trends Fueling Changes in Satellite Applications

There is a great diversity of new technologies, business models and business practices that are working together to make all of these changes in the space and satellite world possible. The number of changes in just the last few years is really almost staggering. It is hard to keep up with the speed of change and where the changes are occurring. Space applications and services are not for those who cannot keep up with almost constant change. Let's count the ways.

There are key changes in the design and technologies now being used in ground systems. These changes are particularly important to the design and operation of low Earth orbit constellations. There are also significant changes in digital encoding systems that allow for more efficient use of the available spectrum. There are new frequency

a

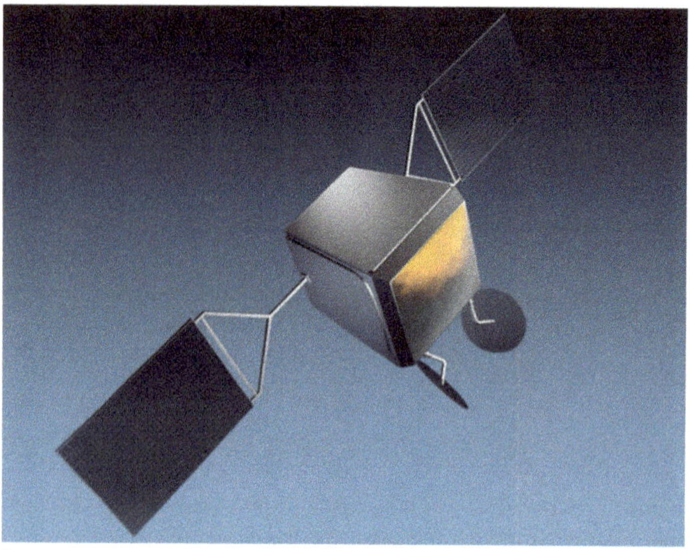

b

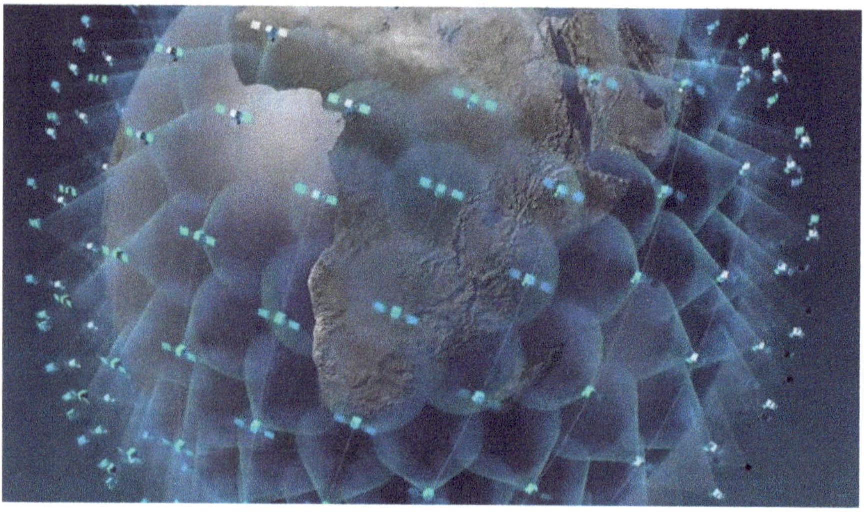

Fig. 1.3 **a.** A mock-up of a small satellite being manufactured by Airbus for the OneWeb constellation. **b.** Representation of the OneWeb satellites deployed as part of a giant 800-satellite orbital constellation. (Graphic courtesy of OneWeb.)

bands that are being made available for some of the new services. There are new ways of manufacturing satellites on assembly lines that allow for faster and lower cost production and quality testing. There are new, lower cost and more capable launch systems. The how and why of these changes and future trends are key elements of this book. The details of these changes are presented in later chapters, but the key highlights are summarized here.

Fig. 1.4 The giant Viasat 2 satellite that has a wingspan of a football field. (Graphic courtesy of Viasat.)

Key Innovations in Ground Systems

The significant factor that Arthur C. Clarke hit on with regard to the use of geosynchronous or geostationary (GEO) satellites was the simplicity they offered on the ground. He explained in 1945 that GEO satellites are essentially stationary 24/7 for 365 days a year. This meant that Earth station antennas on the ground could be continuously pointed to the same location in the skies. That means they could always maintain a link with the satellite above without expensive tracking systems.

As communications satellites became more and more capable with higher power and improved higher gain antennas, smaller and lower cost satellite ground stations became possible. Now there are millions of very small aperture television receive only (TVRO) terminals all over the globe.

The widespread usage of the Internet, with its protocol that was sensitive to satellite transmissions all the way to a GEO satellite and back but involved a quarter of a second delay, created impetus to look at lower orbiting satellite constellations. Further the idea of creating satellite systems for mobile communications and the need to operate with very small satellite handsets also prompted the idea of using low Earth orbit satellites. This was because of not only a shorter transmission delay but also the advantage of less transmission loss.

A satellite signal as it is emitted from a satellite antenna spreads in a circle, and the area of a circle is $A = \pi r^2$. This means that a satellite that is 40 times closer to the ground does not experience

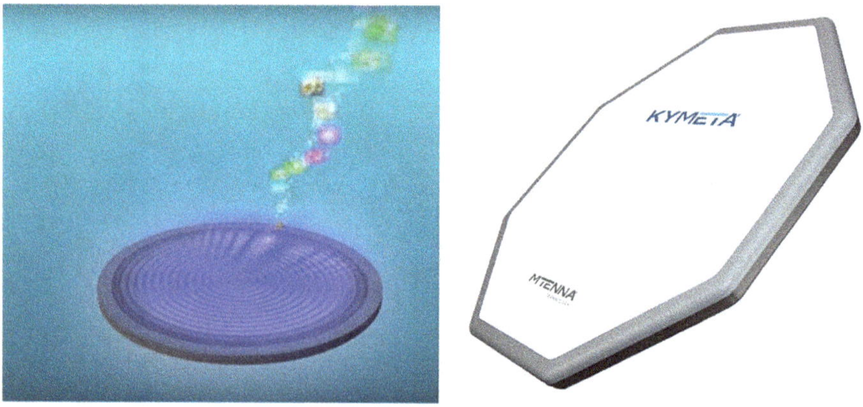

Fig. 1.5 A representation of an electronically generated beam and a Kymeta antenna. (Graphics courtesy of Kymeta.)

40 times less spreading and corresponding weakening of the signal but $(40)^2$. This means that the effective or in power advantage is $(40)^2$ or 1,600. On the other hand, such a satellite has much less coverage of the world's surface, and thus there is a need for more satellites. For global coverage for a constellation that is 40 times closer to the ground you would need perhaps 50 to 60 satellites.

The problem with the low Earth coverage is the need for either a tracking capability to follow the satellite as it passes overhead, or a very low gain or sensitivity to receive the satellite signal. What has changed is that there are now new Earth stations that use what are called meta-materials. These allow new ground antenna systems to generate electronic beams that can track an overhead satellite and both receive and transmit signals to a low Earth orbit satellite as it passes overhead. The leader in the field at this time is a new company called Kymeta that has perfected this technology and is now commercially providing these new types of ground stations that make these new low Earth orbit satellite constellations much more viable. This company has the further cachet of being backed by Microsoft co-founder Bill Gates [6]. (See Fig. 1.5.)

There are other companies, such as Phasor and C-Sat, that are also developing competitive ground antennas [7].

These advances in ground systems and in handheld units based on highly refined application-specific integrated circuits (ASICs) are perhaps as crucial, if not more so, to the advancement in satellite communications services in the years ahead.

Closely linked to this area of development is the creation of new capabilities to support satellite service to Earth Stations on Mobile Platforms (ESOMPs) and Earth Stations in Motion (ESIM). These, too, are now being equipped to operate using electronic beams with continuous pointing capabilities that allow these mobile systems to track satellites in low Earth orbit and to achieve higher gain links with satellites that are constantly moving in space and at the receiving location.

Advanced Coding Systems

Of all the advances in satellite communications, the one that has allowed the biggest increase in efficiency has been the use of digital encoding. The use of advanced COders/DECoders (CODECs) have offered the largest gains as measured by bits transmitted per hertz. The typical digital transmission a decade ago had an efficiency rating of 1 bit/Hz. Today the transmission capabilities, especially with television and high definition television systems, are in the range of 4 bits/Hz up to even 7 bits/Hz. This is like taking a car that had a mileage rating of 30 miles per gallon or 12 km/L up to a spectacular level of 210 miles per gallon or perhaps 84 km/L. Major advances in Trellis coding, Turbo-coding, etc., have allowed tremendous advances. The world of fiber optic communications that has access to the vast amount of spectrum available in the light wave bands is a key to future progress. The details of these advances are provided in Chapter 2 of this book.

New Frequency Efficiencies and Spectrum Allocations

The field of satellite communications is essentially driven by three basic parameters. These are power, radio frequency spectrum and complexity. And the meaning of complexity in this case essentially boils down to digital coding concepts. When one has exhausted the available power that is available for transmission to or from a satellite, and then devised and used the most efficient coding system over a satellite link, the only way to expand one's satellite service capabilities is to find better ways to reuse the available radio frequencies or get new orbital frequency allotments. Such new allotments today would for the foreseeable future come in the form of higher radio frequency spectrum bands. This push up to higher and higher frequencies is the usual way to proceed, since the lower spectrum bands are completely used up by other forms of intensive radio frequency usage.

The new options for satellite communications planners involve the so-called Q/V-bands, the W-bands and even the E-bands. None of these potential newer bands in the millimeter wave band is particularly attractive in that they all involve intense problems of rain and precipitation attenuation, or what is called simply "rain fade." This means that during rain, snow or fog conditions the radio wave transmission could in effect be 'bent,' since the wavelength of the radio wave and size of raindrops are close enough in size that the raindrops act as a sort of lens to distort the signal.

And this is not all when it comes to problems of using the millimeter-sized wavelengths. The frequencies that are being modulated are very high indeed. The equipment needed to operate at these frequencies is difficult to design and build and thus the cost of the radio equipment – both for the satellites and the ground equipment – is today expensive and will likely remain so for a number of years to come. One will need to design satellite systems to be clever, using a number of tricks to cope with heavy rain by reserving power margin and expanded time slots that can be used when rain fade occurs.

At this stage of development the main strategy that satellite operators and manufacturers are inclined to use is to find better ways to reuse frequencies in the microwave and lower millimeter wave bands such as 30Ghz/20 GHz (i.e., Ka bands and the 48/38 GHz – Q/V – bands) rather than migrating to the W and E band frequencies at 60 GHz and higher. The other option still remains, which is to seek even more efficient encoding and digital compression techniques to send more information per hertz.

Of course, the tendency is to push in all of these directions in order to find the best combination of technical advances that give the most bang for the buck; this is to say the most productive and cost-effective solution. The famed baseball catcher and quipster Yogi Berra once said: "When you come to a fork in the road, take it." The satellite industry is indeed pursuing all the forks in the road.

New Manufacturing and Quality Assurance Capabilities

There have been some true breakthrough capabilities in the space applications industry in the past decade with regard to the design, manufacture and quality assurance testing. This has let the cost of designing, manufacturing, and quality assurance testing of satellites to drop significantly. In many instances the designing in of quality into the manufacturing processes has allowed the cost of quality assurance testing to drop dramatically.

New Launch System Efficiencies

For over a half century, the largest barrier to cost efficiencies in the space industry has been the relatively static cost of launch vehicle services. The developing of reliable launchers in China and India and other countries started a downward trend in launch costs. Then a number of new commercial launch companies began to develop new designs and new approaches to developing and deploying rocket systems. Some of these efforts involved developing carrier systems to launch space systems or rocket launchers from high altitude, and a few involved new fuel systems and greater launch safety. Other systems involved new modes of manufacture – even 3D printing, or additive manufacture of rocket engines.

Perhaps most significantly Elon Musk's SpaceX has sought to develop rocket systems that might be reused rather than be considered expendable vehicles. Musk has not only demonstrated that rocket systems can be successfully reused but they can be landed with precision at preset points. His objective is to develop rocket systems that can be reused twenty to twenty-five times with a minimum of refurbishment [8].

And Jeff Bezo's Blue Origin company has indicated that their rocket systems will follow a very parallel path and that his rocket program should be able to support space tourist flights as well [9].

The bottom line is that there is a new array of approaches to launching systems into space that are significantly lower in cost. These new developments are a part of the Space 2.0 revolution.

The Structure of This Book

This book is an introduction to the world of NewSpace and the many changes that are occurring in space industries. The technologies that are being used today owe their heritage as much to companies of Silicon Valley and their emulators around the world than to the traditional aerospace companies and the space agencies. These technologies are thus new, different and in some cases revolutionary. The type of space industries and the new space applications are also different, as we are seeing the spread of new ways to derive value from space services. This then has led to new business models and new approaches to the regulation and governmental oversight of the various types of space industries.

As mentioned earlier, these many changes to the world of space applications are sometimes referred to as either Space 2.0 or NewSpace. This book seeks to cover all of these changes in the world of space and to explain how they have occurred. This first chapter has sought to introduce how and why the space industry has changed in a fundamental way in the past decade and why the revolution will continue for many years to come. The rest of the book seeks to describe these many changes into a greater detail and explain key components of them so that one can see how the various parts form a new whole that is at once exciting and able to offer new opportunities to investors and global consumers. It is also a time of tremendous challenge to engineers, space entrepreneurs and space business leaders, as well as those involved in the regulation and oversight of this totally new era of the space enterprise.

To summarize the contents of the chapters:

Chapter One. The World of Space in Flux. The world of space applications is today in enormous flux. This chapter seeks to identify the most significant changes in space systems technology and changes in market applications that are bringing new participants into the field, and the most vital forces that are reshaping the most important changes in today's space markets. The changes include not only new technologies and applications but new vulnerabilities and concerns that will reshape the future.

Chapter Two: How Satellite Communications Systems Are Changing. The largest, most well-established and perhaps the most profitable space industry today is the satellite communications and network services industry. Today, this largest space industry is changing because almost everything has changed. There are new ways to design, manufacture, and test the satellites plus new types of Earth stations and new launch options. Even the deployment configurations in orbit have changed as well as the latest encoding concepts. Most industries can change because one of their dimensions has changed. Rarely do five major dimensions change at once, and the future promises even greater change as the markets adopt new technologies.

Chapter Three: Key Trends in Remote-Sensing Satellite Systems Systems. The world of remote sensing has seen the same chaotic change as has been experienced in satellite communications. Small satellite systems have not only lowered the cost of building and testing remote sensing satellites but have also dramatically decreased launch costs. Since remote-sensing satellites do not require high gain and larger aperture antennas, the small satellite revolution came earlier to the world of Earth observation, and we now even see remote-sensing satellites capable of delivering hyper spectral sensing services.

Chapter Four: The Growth and Expansion of Precision Navigation and Timing. The big question in satellite applications is when will we see a major revolution in the world of satellite navigation and precise timing satellites? The high cost of atomic clocks has to date not allowed for major breakthroughs in the cost of space systems, but remarkable progress in the decreased cost of user units. The fact that today's smart phones are universally equipped with global navigation satellite service (so-called sat-nav capability) has turned this service into the second largest space application. If it is ever possible to reduce the cost of the atomic clocks it will also create a revolution in space-based systems as well. Even without such a breakthrough, the applications are expanding rapidly, and the number of systems being deployed around the world is also multiplying.

Chapter Five: The New Capabilities of Weather Satellites. Weather satellite systems are still largely governmental operations, but the technology and the capabilities continue to expand. New systems with lightning trackers can pinpoint where storms are most intense and their immediate tracking vectors. There are also new capabilities to track and monitor climate change and also to provide much closer tracking of solar storms, radiation flares and coronal mass ejections. These capabilities are critical to protecting vital resources such as satellites, electrical grid systems and vulnerable pipeline control systems.

Chapter Six: New Uses of the Protozone. One of the most important new areas of space applications are not in space at all but involve the use of the area above commercial air space and below orbital space. This area, sometimes called sub-space, near-space or the protozone, is now an area of intense study with many new applications either under study, in experimental use or actual use. Here there are a number of competing technologies and applications that give rise to many regulatory and safety issues.

Chapter Seven: On-Orbit Servicing and Repurposing of Defunct Satellites. Another space application that has been the subject of experiment and testing for years involves on-orbit servicing. Finally, this technology has reached a level of maturity that it now seems ready to reach liftoff potential. After years of speculation by space experts, there are now beginning to be actual commercial ventures seeking to provide a wide range of in-orbit servicing, retrofit, satellite repositioning and even in-orbit construction and processing.

Chapter Eight: Space-Based Solar Power Satellite Systems. For some fifty years there has been speculation about the potential of solar power satellites and their potential to bring clean energy to the world on a 24/7 basis. Breakthroughs in space technology and ground rectenna system design suggest that these types of systems might be able to become financially viable in the not too distant future.

Chapter Nine: Space Weapons, the Threat of War in Space and Planetary Defense. The divide between civilian commercial activities and defense-related applications is a part of this discussion. The possibility of creating 'space forces' and the possibility of conflict in space is one of the major issues that could alter or even reverse the rapid rise of new commercial space applications. Cosmic hazards and manmade nuclear denotations that give rise to electromagnetic pulses are just some of the concerns that space operators must consider, along with frequency interference and orbital space debris.

Chapter Ten: Trends in Chemical Rocket Systems and New Approaches to Launching Satellites. Many of the changes in today's space applications comes from the development of new more cost-effective and flexible launch systems that are more reliable, cost-effective and versatile than launch systems of the past. Many suggest that the whole new Space 2.0 revolution is due to the changes in commercial launch capabilities and their ability to support small satellite launches and other new types of space deployment requirements.

Chapter Eleven: The Longer Term Future of Launch and Propulsion Systems. The revolution in launcher systems and better ways to get space systems in orbit and operate suborbital flights for space systems is, in some ways, just beginning. There are a range of new space technologies that are under research and development. Some of these systems seem on the brink of moving forward, such as electric propulsion systems, while others may still be some ways away from operational service. The status of these technologies and their potential for the future will be examined and explained.

Chapter Twelve: Spaceplanes, Space Tourism and Private Space Habitats. Some of the big new areas of commercial space development that have perhaps made the public most aware of the Space 2.0 revolution have been the developments in the area of commercial spaceplanes and commercial space habitats. The Burt Rutan and Paul Allen spaceplane ride into space in 2004 in order to win the X-Prize in 2004 was seen as the true kick off for these private ventures to fly into space. Billionaires Richard Branson and Jeff Bezos are now racing to be the first to offer civilian rides to space on spaceplanes that make suborbital flights. Meanwhile billionaire Robert Bigelow is now testing an inflatable habitat on the International Space Station as a prelude to offering private citizens the opportunity to stay in space on a private space station.

Chapter Thirteen: Space 2.0 Economic, Business and Regulatory Issues. The future of space applications does not depend on technology or developing new business markets alone. Rather the key to the success of these new space applications may well depend on new regulatory agreements on how to provide legal or administrative guidelines for the operation of these activities. Some of the top issues currently under debate at the national, regional and global level will be discussed.

Chapter Fourteen: The Way Forward. This book will end with summarizing the top trends in the world of space applications that are critical to the future success of commercial space and the potential of space to become one of the leading areas of expanded commercial applications in the decade ahead.

Conclusions

In these fourteen chapters, a brief history of the many space applications and their technological development will be provided. This will be accompanied by a current status review of the many changes that are occurring in ground systems and in the financing and start-up of new ventures in the field. Perhaps most significantly this book will touch on the very latest innovations in technology and space markets. This will allow the tracing of the latest trends to see how they will allow a continued rapid expansion of space systems and services within the next decade. Thus, the rapid evolution of space technologies, services and markets will be reviewed, and key trends for the future explained. In the coming decade, it is quite likely that the multi-billion dollar space industry will make great strides toward becoming a trillion dollar enterprise if current technological, business and regulatory challenges are successfully overcome.

References

1. Frischauf, N., Horn, R., Kauerhoff, T.I., Wittig, M., Bauman, I., Pellander, E., Koudlka, O.: New space: new business models at the interface of space and digital economy: chances in an interconnected world. New Space. **6**(2), 135–146 (2018). https://doi.org/10.1089/space.2017.0028
2. Levy, S.: Jeff Bezos wants us to leave earth for good - inside Blue Origin's mission to an alien future. Wired, **55** (2018). https://www.wired.com/story/jeff-bezos-blue-origin/
3. Pelton, J.N., Marshall, P.: Megacrunch: Ten Survival Strategies for 21st Century Challenges. The Emerald Planet, Washington, D.C. (2010)
4. Berry, E.: India launches 104 satellites from a single rocket, ramping up a space race. New York Times. https://www.nytimes.com/2017/02/15/world/asia/india-satellites-rocket.html (15 Feb 2017). Accessed 17 May 2018
5. Pelton, J.N., Madry, S., Camacho-Lara, S.: Introduction to satellite communications. In: Handbook of Satellite Applications, 2nd edn. Springer, Basel (2017)
6. Why Kymeta. https://www.kymetacorp.com/why-kymeta-connectivity/. Accessed 18 May 2018
7. Phasor successfully completes broadband satellite transmit tests from a moving platform. http://www.phasorsolutions.com/news-1/phasor-successfully-completes-broadband-satellite-transmit-tests-from-a-moving-platform (27 Sept 2016)
8. Wall, M.: Elon Musk says SpaceX will reuse a rocket within 24 hours in 2019. Space.com. https://www.space.com/40581-spacex-reusable-rocket-goal-elon-musk.html (15 May 2018)
9. Wall, M.: Jeff Bezos' Blue Origin to launch reusable rocket again Saturday. Space.com. https://www.space.com/32451-blue-origin-reusable-rocket-launch.html (1 Apr 2016)

2 How Satellite Communications Systems Are Changing

Introduction

The field of satellite communications is highly competitive and rapidly growing. Today, this sector of the global space industry market, including defense and commercial satellite communications, represents annual revenues of nearly $200 billion out of total revenues of over $350 billion [1].

For a half century there has been a predominant pattern of technological development in the satellite communications field. We have been designing, building and launching larger, more massive, higher-gain satellites that can operate with smaller and less expensive ground systems. Today there are literally millions of ground satellite user terminals and antennae. These units that are sometimes as small as hand-held transceivers are accessing and using communications satellites in some 200 countries and territories around the world.

This pattern of development has existed for a half century, but it is now suddenly changing. Indeed there are a number of important new innovations running in several different directions. The final outcome with regard to these various conflicting innovations is far from clear. A recap of the rise of satellite communications and an analysis of the many new directions are addressed in this key chapter, as the reinvention of the satellite communications industry is explored and key trends analyzed.

The Rise of Conventional Communications Satellites from the 1960s to the 1990s

This is a condensed history of the development of satellites. A more complete history can be found in the author's much larger work, *Handbook of Satellite Applications* (Springer, Second Edition, 2017). We are at a point of diverging into new competitive streams of technology and potentially new global markets. This is a high stakes gamble not only for satellite communications but other NewSpace applications that

J. N. Pelton, *Space 2.0*, Astronautical Engineering,
https://doi.org/10.1007/978-3-030-15281-9_2

depend on supportive financial markets to fuel new innovation and the rise of new space systems.

In the 1960s, two major technical conclusions were reached about how to offer viable satellite communications services. Firstly, the big balloon experimental satellite ECHO, launched in 1960, confirmed that using a passive reflective surface as a satellite to bounce electronic signals off of was much too inefficient to be economically viable for commercial service. Secondly, Syncom 2 and 3 in 1963 confirmed that one could successfully place a satellite in geosynchronous (GEO) orbit and operate from this special type of very high orbit almost a tenth of the way out to the Moon. This unique orbit allowed a fixed Earth station on the ground to not require expensive and rapid tracking mechanisms.

The first satellite launched for commercial satellite communications was Early Bird (or Intelsat I). This was a so-called GEO satellite that was an expanded version of the Syncom satellites built by Hughes Aircraft Company – now morphed into the Boeing Corporation. This small beach ball-sized satellite was able to provide only 240 telephone circuits or one low-quality black and white television channel. As the first commercial communication satellite it was power limited, had a low gain squinted beam antenna, lacked the ability to point precisely back to Earth, and was limited to a single use of the C-band spectrum. These many limitations in satellite power and performance required these giant ground stations to be tremendously expensive multi-million- dollar facilities. In addition, these Earth stations had very large aperture antennas equipped with very high performance low noise amplifiers. They also required an extensive round the clock staff of 40 to 60 people.

In the years that followed the satellites grew in size and capability. They became more complex, more capable, and were equipped with higher power. These increasingly large satellites developed the ability to reuse frequencies not only in the C-band but in other higher frequency bands as well. Over time, commercial satellites moved upward in frequencies to include the Ku-band, the Ka-band and most recently in the Q/V bands as well. These communications satellites were for the most part deployed in the geosynchronous orbit in order to allow ground antenna systems to stay pointed to the satellite above rather than requiring constant tracking of the satellite as it moved over the horizon.

All of these innovations in satellite design that have occurred over a period of decades allowed the satellites to become a thousand times more capable and send thousands of times more telephone circuits, data and television channels. The main gains were:

- Much higher power on board the satellites (e.g., large solar arrays and bigger batteries).
- Larger high-gain aperture antennas on board the satellites that could be constantly pointed toward Earth and also were equipped for precise antenna beam pointing.
- Polarization techniques that allowed reuse of the available spectrum.
- Complex feed systems that allow many beams to be generated from high-gain reflectors. This allowed even more reuse of frequencies and focusing of tightly formed spot beams to limit beam power spreading and thus allow concentrated power to specific locations.

- Access to broader bands of spectrum in available higher frequencies – and much more.

All of this effort concentrated on making the satellites more powerful, having access to more and more RF spectrum, and also adding to the complexity of signals through the encoding of digital communications signals. These digital complexity techniques paid off in the efficiency of information transmission via the available spectra. All of these many gains also meant cost reductions, downsizing and simplification of the ground antennas for users.

Over the decades we saw more and more powerful satellites and more usable spectra both through more intensive frequency reuse and use of more spectrum bands. When these gains were combined there was the equivalent of hundreds of times more radio frequency spectrum that could be used for satellite communications around the globe.

The greatest gain in efficiency of satellite operations came via complex digital encoding that allowed much more information to be sent through the available spectrum. Virtually all of these efficiency gains in satellite operations and design allowed the ground antennas to become smaller and lower in cost and then even fully automated in their operation. There was no longer a need to staff Earth stations. The advent of digital satellite communications brought the greatest efficiency gains through the use of encoding to send more information or data per Hertz of bandwidth. The biggest barrier to satellite communications efficiency gains throughout this period was the lack of cost reductions for satellite launches that remained stubbornly resistant to new cost efficiencies.

These decades-long advances to develop more efficient and cost-effective satellites also enabled the reduced cost and size of ground systems. This trend became known as "technological inversion." This meant more and more complex and powerful satellites in the sky with more access to new radio frequencies and more spectrum via frequency reuse allowed smaller and lower cost ground antenna systems. In short, the satellites were bigger, more powerful and more costly, but this enabled smaller and less costly units on the ground. (See Fig. 2.1a and b to see

a **b**

Fig. 2.1 **a.** Tiny Early Bird (Intelsat 1) and **b.** Gigantic Intelsat Epic satellite in 2018. (Graphics courtesy of Intelsat.)

enormously increased satellite power, antenna gain, and throughput capabilities.)

Ground systems have shrunk in size but enormously increased in numbers from one giant Earth station per country to millions of small satellite ground systems spread over the globe. The satellites grew in cost by ten times and then even a hundred times more in order to build and launch. Yet this allowed dramatic decreases in the cost of the user terminals and the spread of low-cost antennas all around the world. The overall system costs remained in balance between the cost of the space-based systems and the systems on the ground.

These ground systems indeed shrank to very small aperture antennas. Instead of costing millions of dollars, the costs of ground systems shrank to only thousands or then even hundreds of dollars for receive-only television terminals. The ultimate shrinkage has now led to hand-held units used for mobile communications and the very smallest receive-only satellite television dishes that in some cases are as small as cereal bowls. These small dishes are nevertheless capable of receiving multiple television channels from the highest powered direct broadcast satellites. The predominant trend of technology inversion from 1965 through the 1990s is shown in Fig. 2.2.

This trend that allowed the ground stations to shrink from 30-meter-high gain antennas down to VSATs and now even hand-held units is shown in Figs. 2.3a, 2.3b and 2.3c.

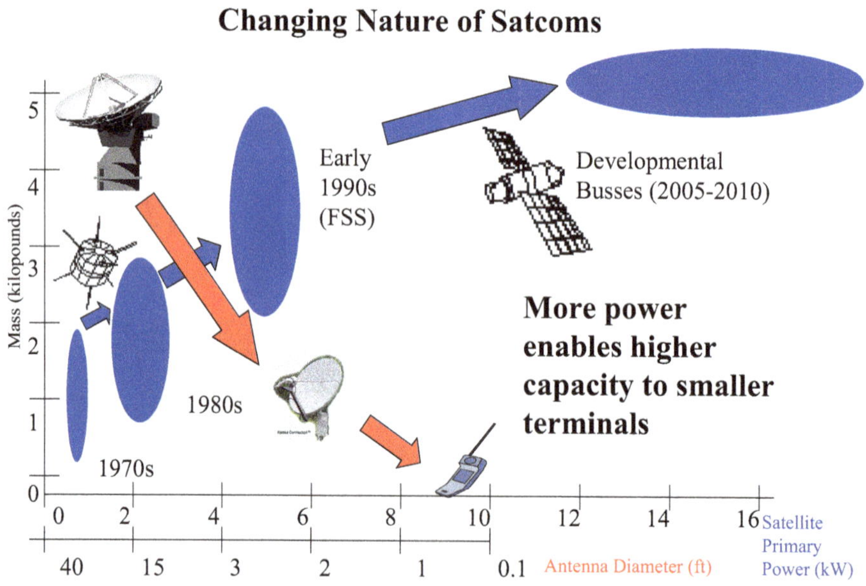

Fig. 2.2 Early 1960s small satellites in GEO orbit have grown to powerful satellites with large multi-beam antennas allowing ground antennas to shrink in cost and size. (Graphic provided by the author.)

a b c

Fig. 2.3 **a.** A 1960s giant Earth station. (Illustration courtesy of Comara.) **b.** VSAT small terminal from the 1990s (Illustration courtesy of Hughes). **c.** Sat phone of today. (Picture courtesy of Iridium.)

The Rise of New Space Communications Systems from the 1990s to the Present

In the 1990s, however, several things began to change to allow a new pattern of development for satellite communications. This new pattern perhaps first began when some satellite designers began to question the mainline concepts of putting virtually all of the communications satellites in the geosynchronous (or Clarke) orbit that is 35,850 kilometers (22,230 miles) out in space. As already noted, this is the special orbit where ground stations do not have to actively track what is essentially seen as a satellite hovering above in the sky.

The problem was that this orbit is way out in space, almost a tenth of the way out to the Moon, and this very high altitude orbit comes with penalties. Communications satellite engineers, who were looking for a new approach, explained that while it was useful not to constantly track the satellite, this very long transmission path results in what is called by satellite communications experts "path loss." Further there is also a time delay or latency that represents a problem for voice and data networking services. The further the satellite transmission has to travel, the greater the delay. Even at the speed of light, there is a quarter-second delay from the ground to the satellite and back down. With the return link for someone talking at the other end, this can mean a half-second delay and is a problem with normal telephone conversation. Even a quarter-second delay is a problem in computer networking.

Engineers noted that if the satellites were 40 times closer, the effective power advantage, due to less path loss or reduced beam spreading was 1,600 times greater. This is because antenna transmission spreads out in the form of a circle (i.e., the area of a circle is $A = \pi r^2$). This meant that the loss in signal strength was equivalent to the square of the distance represented by the satellite orbiting above Earth. They also noted that if you wanted to provide mobile communications to ground systems they would need to be moving in any event to receive the satellite signal. Thus these satellite engineers argued in favor of a network of low Earth orbit satellites. They conceded that because the satellites would be much closer to Earth there would need to be constellations of at least 50 or so to blanket the globe effectively at an altitude as low as perhaps 500 to 800 km (or about 310 to 500 miles) above Earth's surface.

This led to the development of several new satellite communications systems that could provide mobile communications satellite services. These new systems departed from the usual practice of using the predominant geosynchronous orbit. Innovators came up with the idea of using a constellation of satellites in low Earth orbit to provide global mobile services. Those systems that were actually deployed in the late 1990s included the Iridium Satellite System and the Global Star Satellite System.

Another system known as ICO, that was a spin-off of the INMARSAT system for maritime communications, was never deployed but followed Iridium and Globalstar into bankruptcy. These systems were designed for voice land mobile communications and engineered to connect to hand-held units. In addition, there was the OrbComm satellite system, which was designed to provide store and forward data communications or machine-to-machine (M2M0) service. Another system named GEOstar that used different frequencies and only allowed short messaging was also deployed in a low Earth orbit constellation during this time of innovation.

All of these innovative systems, for several reasons, initially failed and the companies went into bankruptcy.

In the case of Globalstar and Iridium, there were several factors. These included the high cost of the voice-based land mobile satellites services. The cost of the satellites, ground systems and user terminals, the regulatory constraints created by national tariffing policies on landing rights and user terminals ultimately ended up being more expensive than had been first estimated. Charges ended up being quite steep, i.e., between a \$1 a minute to even \$10 a minute. Most significant was the fact that terrestrial cellular services had greatly advanced in coverage and power margins during the time that Globalstar and Iridium systems were being designed, manufactured and launched. And also the satellite hand-held units were large in comparison to the cell-phones that were being manufactured some seven or eight years later.

The satellite phones were sometimes called "bricks." Perhaps the important fact was that these LEO mobile satellite systems did not have enough power margin, so that they typically did not operate within houses or buildings, and even in cars they did not operate with a sufficient degree of reliability. These factors all contributed to a lack of significant growth of the market for the satellite phone service. The millions of users that market analysts had projected did not develop. The customer base was instead in the thousands. The result was a series of consequent bankruptcies for Iridium, Globalstar and ICO.

Likewise there was also a lack of market penetration by the store-and-forward data services using M2M messaging. Thus these other satellite systems also failed. Orbcomm went to the bankruptcy court as a financial loss and Geostar did not survive as a service provider. Eventually the Globalstar, Iridium, and Orbcomm systems were reorganized and under new management and ownership did re-emerge and are now still providing service through second-generation spacecraft, but the initial damage to in terms of market support had been done. The markets were skeptical of new satellite communication constellations in low Earth orbit.

There was in the late 1990s yet another proposed system to be deployed in low Earth orbit that was described as a mega-LEO system. This was a satellite system proposed to provide a broadband Internet in the sky. It envisioned the provision of broadband services for fixed satellite services that would have been in competition with organizations such as Intelsat, Eutelsat and other such service providers. In this case the proposal was for launching nearly 1,000 satellites plus spares in a giant LEO constellation. This system design was highly innovative and was envisioned as being able to provide high data rate services using Ka-band (30 GHz/20GHz) spectrum. The concept was to design, manufacture and launch these satellites on a mass-produced and highly efficient basis. The plan was to benefit from economies of scale in production and qualification testing, unlike the limited production levels that had generally been used for GEO orbital satellite networks in the past. This system, known as Teledesic, had the additional feature of being backed by entrepreneur Bill Gates. This planned system went into bankruptcy before any of its 980-plus spare satellites were deployed. But then a decade passed.

Over time, Iridium, Globalstar and Orbcomm all came back out of bankruptcy and thus the feasibility of the use of low Earth orbit constellations did begin to be taken seriously again in financial and business markets. Further groups, such as the Surrey Space Centre Ltd., was producing small satellites, known as "Surrey sats." at quite low costs. Others such as Skybox and Planet Labs were producing small sats using off the shelf materials and deploying new systems at very low cost for remote sensing. (This is a subject that will be addressed in the next chapter.)

On top of everything, additive manufacturing or 3D printing was starting to show how key components of satellites could be manufactured at very low costs. Collectively all of these factors combined to produce new interest in the idea of how low Earth orbit constellations might be designed to create new satellite systems to provide telecommunications and networking services to underserved parts of the world.

However, it was during this period beginning around 2010 that many different innovations sprang up at once and created new synergies.

Space 2.0 Comes to Satellite Communications

The world of satellite communications since 2010 has been turned upside down and some would say almost reinvented. Satellite engineers have now designed, built and launched high throughput satellites (HTS) that are able to operate at truly prodigious speeds for space systems. Throughput speeds of 140 gigabits per second have been achieved with Viasat 1 and 2, with Intelsat Epic satellites and Hughes Network Systems Jupiter satellites not far behind. These high throughput satellites (HTS) have ten to fifty times higher throughput rates than conventional satellites of only a few years ago and have continued the conventional trend lines of finding more ways to reuse RF spectrum, adding more power and exploiting the capabilities provided by the latest in digital encoding technology. These satellites with their greater power can link to even lower cost ground stations.

Other satellite designers, however, are moving in the direction of small sat constellations flying in low Earth orbit that would deploy a very large number of satellites.

These new ventures are finding ways to design and build small satellites for large-scale constellations that can be built on assembly lines at high speeds and use additive manufacturing to build key components at lower cost and with higher reliability and exactness. They are not as small as cube satellites, because antenna diameters have to be larger to achieve needed gain, and there is also a need for higher power. Yet these small satellites with a mass typically ranging from 200 to 400 kilograms (440 to 880 lbs.) are ten to fifty times smaller than giant high throughput satellites (HTS) that Viasat, Intelsat, Inmarsat, SES or Echostar/Hughes Network Systems are now placing into service. There are of course many more satellites in these constellations than in GEO-based systems, but it is much easier to launch smaller satellites, especially to low Earth orbit.

In mass production these small satellites are much lower in cost than their big brothers, even after taking their relative mass into account. Perhaps their biggest advantage in terms of performance is due to the fact that they are much closer to Earth. This gives the advantage of much lower path loss and perhaps even more importantly up to 40 times less latency to support more effectively either voice or networking services.

This new approach of deploying satellites in non-geosynchronous orbits started with the O3b system (standing for Other Three Billion people in the underserved world). The O3b network deployed 12 satellites in medium Earth orbit (MEO) initially and then added 6 more, and in the latest filings dozens more are planned for launch. Gregg Wyler, the entrepreneur who started O3b, has been focused on finding new ways to provide communications and networking services to the developing world for a couple of decades and has moved on to an even more ambitious venture. He has now sold out his interest in O3b to SES of Luxembourg, and has moved on to acquire a new group of partners and raised the capital to launch the very ambitious OneWeb satellite constellation in low Earth orbit (LEO).

This system is currently just starting to be launched, and in the next few years through 2020 or 2021 will deploy about 800 satellites, including spares to provide networking services in new ways throughout the developing world. Thus OneWeb is particularly designed to provide coverage and Internet-optimized services in areas such as Africa, Asia, the Middle East, South and Central America and the Caribbean plus the South Pacific islands. But all is not smooth sailing; the cost per satellites for the OneWeb system have increased, and the overall system has not been financed. Further the cost of flat-panel ground systems that can electronically track the fast-moving LEO satellites are currently around $30,000 apiece from suppliers such as Kymeta. In short the cost of the satellites and ground antennas for LEO constellation systems for communications and networking services are higher than were first estimated. Further it is not clear that the manufacturers of the new flat-panel antennas could possibly meet the huge expected demand.

According to the last figures presented by Northern Sky Research there are now 25,000 constellation satellites filed for launch. There are thus several quite serious challenges here that could be a show stopper for many of these constellations. These problems are thus: (a) cost of manufacture of many of these commercial small satellites might be higher than first estimated; (b) the cost of flat-panel antennas capable of tracking LEO satellites may stay higher than is needed to support service in rural areas; (c) the supply of tracking ground systems may be greatly inadequate to meet the huge demand that will be needed to provide the connectivity for actual users; and (d) there may be inadequate launch capability for all of these satellites at least on the schedule that the many new small satellite system operators would hope to achieve. The good news is that many new LEO satellite constellations will be deployed and provide important new services. The bad news is that for the above four reasons a number of the filed systems will fail. And that is not all the problems to be solved. New regulations to control the proliferation of satellites, minimize radio frequency interference, cope with orbital space debris or limit pollution and particulates in the stratosphere could create new regulatory hurdles as well. What is clear is that the next ten years will be a time of great turbulence [2].

Table 2.1 provides a listing of the many of the filings that have been announced and registered with regulatory authorities. One can see from this chart the many diverse plans for proposed new entrants seeking to build and launch what are typically designed as

Table 2.1 Listing of some of the proposed small sat constellations for communications. (Listings were prepared by the author.)

State	Constellation	# of Sats	Radio Frequency Bands
Canada	CANPOL-2	72	LEO and highly elliptical Earth orbit in VHF-, UHF-, X-, and Ka-bands
Canada	Telesat Constellation	117 satellites plus spares	LEO in Ka-band
Canada	COMSTELLATION	Nearly 800 Satellites	LEO in Ka-band
France	Thales Group's MCSat	between 800 and 4000	LEO, MEO, and highly elliptical Earth orbit in Ku- and Ka-bands
Liechtenstein	3ECOM- 1	264	Ku- and Ka-bands
Norway	ASK-1	10	Highly elliptical Earth orbit in X-, Ku-, and Ka-bands
U.K.	L5 (OneWeb)	750 plus spares	Ku- and Ka-bands
U. S.	Boeing	1396-2956	V-band in 1200 km orbit
U. S.	SpaceX	Up to 4000	Ku & Ka band
U. S.	SpaceX	7500 plus	V-band
U. S.	Leosat	About 80	Ka-band

small sat constellations to be deployed in low Earth orbit. These proposed systems, however, must be viewed with some skepticism, based on past history.

Back in the 1990s there were 17 filings to launch a number of new Ka-band satellite systems. These were all submitted to the U. S. Federal Communications Commission. Of those filings, which most notably started with the Teledesic system filing, only the Ka-band satellite system, originally known as Wild Blue, was ultimately fully deployed as filed. And this was a GEO orbit system and not a new-type LEO constellation.

Key to these newest small sat constellation projects going forward are two additional factors that extend beyond the idea of achieving low-cost mass production and new quality assurance testing of high volume production spacecraft. One important factor is that of much lower cost launch systems, including reusable launchers, and the other is a key new and almost revolutionary development in Earth station technology. Currently SpaceX and Blue Origin are leading in the development of new reusable launchers that promise to lower launch costs significantly. It appears that Stratolaunch, which is backed by Paul Allen and his Vulcan Inc, will likely soon provide yet another option to provide new lower cost launch options. For small sat launches, Launcher One by Virgin Galactic (Sir Richard Branson's company) and Vector One are yet other companies that are bringing new lower cost launch services to the market.

These various efforts to reinvent the satellite launching industry will be addressed in a later chapter. It is only important to note here the significant fact that if launch costs could be cut in half – or more – then these systems become much more affordable to place in service and resupply if there is a satellite failure.

Perhaps the biggest impetus for LEO constellations, however, comes from the new type of satellite Earth stations that use electronic beams that form as a result e of meta-materials in their design. This allows the design and manufacture of flat antenna systems that can electronically track a low Earth orbit satellite as to moves over the horizon in about 7 or 8 minutes of time.

This electronic tracking via a reasonably low cost Earth station is truly a game changer. It allows the ground segment part of these satellite constellations to become affordable and tracking systems to be more reliable. The Kymeta Earth station company that is now producing these new type ground systems represents a key part of this new revolution in the satellite communications business. Again the interesting angle is that Bill Gates, the co-founder of Microsoft, is a key investor in Kymeta [3]. Another company, known as Phasor Solutions, is also now coming to market with new satellite antennas that have electronic beam tracking systems that will seek to compete with Kymeta [4].

And the various listings of small sat constellations provided in Table 2.1 are far from a complete compilation. The number of additional filings for new small sat constellations or supplemental additions to systems already filed simply keeps growing. Below is a listing of additional filings received by the U. S. FCC for additional systems, including additions to the O3b MEO satellite constellation and a new MEO constellation by Viasat.

- Audacy: 3 MEO relays to communicate with LEO spacecraft. (SATLOA2016111500117)

- Karousel: 12 IGSO satellites for video (SATLOA2016111500113)
- Kepler MULTUS: 2-140 LEO nano-sats-M2M communication (SATLOI 2016111500114)
- O3b: Amendment to add another 40 satellites (SATAMD2016111500116)
- SpaceX: With its huge number of satellites has its own thread (SATLOA2016111500118) http://forum.nasaspaceflight.com/index.php?topic=41634.0
- Space Norway: 2 satellites in high-inclination 16-hour orbit (SATLOI 2016111500111)
- Boeing: 60 IGSO (This is separate from small sat system they previously filed) (SATLOA2016111500109)
- Theia: 112 for remote sensing (SATLOA2016111500121)
- Viasat: 24 in polar MEO (SATLOI 2016111500120)

As noted above the combined tally of communications and remote-sensing constellations now filed from countries around the world is around 25,000.

The Promise, the Opportunities, and the Pitfalls

The satellite communications industry today is clearly at a crossroads. It seems likely that there will be a number of clear-cut winners and losers that will emerge over the next five years. The new high throughput satellites are five times or more cost efficient than many of the conventional satellites currently in operation. This is true for systems that provide either fixed satellite services (FSS) or broadcast satellite services (BSS).

The bottom line is that many satellite systems now in operation will potentially be priced out of existing markets. These higher throughput satellites put enormous economic pressure on the less cost-efficient satellites now in orbit and especially those which have not been fully amortized. Another danger is that some satellite systems have been loaded up with heavy debt and are subject to financial pressures to perform in a very highly competitive market.

There are even more questions about the extent to which high throughput satellites in GEO orbit will be in serious competition with many of the planned large-scale LEO constellations. Some argue that the new LEO constellations are largely seeking new markets in currently underserved areas. Thus they are targeted to provide Internet connection in areas where there are currently no telecommunications, data or Internet links in service. At one point Intelsat, the world's largest fixed satellite service provider, was going to become a major investor in the new OneWeb constellation in a deal that was to be financed by Softbank. The basis of the deal was that this merger with OneWeb would feed new businesses into Intelsat and the two systems were largely not in competition. This particular business arrangement that would have led to a $14 billion merger fell through, and thus this proposition was never tested [5].

What is clear is that of the various LEO constellations currently filed, only the LeoSat filing to launch some 80 highly capable satellites has advertised its offering as geared to business enterprise networks as opposed to those largely leveraged to provide new types of networking services to underserved portions of the world. The LeoSat website thus explains its alternative approach to its constellation's proposed services thusly:

> The LeoSat system is being developed in conjunction with Thales Alenia Space, a company with unmatched expertise in designing and manufacturing low Earth orbit constellations. The high-throughput satellites (HTS) will form a mesh network interconnected through laser links, creating an optical backbone in space which is about 1.5 times faster than terrestrial fiber backbones, thus creating a paradigm shift in the use of satellites for data connectivity – rather than a gap filler or last resort where no terrestrial alternative is available [6].

What does seem clear is that the very large number of constellations that have now been proposed seem to require a huge amount of new capital investment for what many market analysts see as largely virgin territory for totally new services. Thus most of the various small sat constellation filings are to put up satellites without an existing market or established revenue stream. Past experience, as shown by Teledesic and the original Iridium, Globalstar, ICO and Orbcomm systems, clearly raises some red flags. There are serious concerns as to whether all of the proposed systems can become financially viable. This seems to be a clear case of technology push driving most of these new satellite filings as opposed to any established or clear-cut market pull for all of these new communications satellite and networking constellations.

Even more to the point it should be noted that the structure of satellite communications revenues are strongly geared toward the direct provision of consumer services in the form of retail sales of direct broadcast entertainment services. The other parts of the industry revenue streams are much more modest. As can be seen in Table 2.2 there are revenues north of $100 billion for consumer services, and fixed and mobile satellite services bring total annual revenues to around $130 billion. In contrast, revenues from satellite manufacturing ($13.9B), launch services ($5.5B), and Earth station sales related to communications satellite services (around $40) totaled around $60 billion in 2016 [7]. These revenue figures do not include figures related to defense communications satellite networks.

What is not clear about all of the new low Earth orbit satellite

Table 2.2 Communications satellite services over a five-year period [7]. (Source is Information Satellite Industry Association, State of the Industry Report, 2017.)

Analysis of revenue streams for commercial communications satellite services					
Year	**2012**	**2013**	**2014**	**2015**	**2016**
Consumer services	**$93.3**	**$98.1B**	**$100.9B**	**$104.2B**	**$104.7B**
Satellite TV	$88.4B	$92.6B	$95.0B	$97.8B	$97.7B
Sat Radio	$3.4B	$3.8B	$4.2B	$4.6B	$5.0B
Sat Broadband	$1.5B	$1.7B	$1.8B	$1.9B	$2.0B
Fixed	**$16.4B**	**$16.4B**	**$17.1B**	**$17.9B**	**$17.3B**
Transponders	$11.8B	$11.8B	$12.3B	$12.4B	$11.2B
Managed Service	$4.6B	$4.6B	$4.8B	$5.5B	$6.2B
Mobile	**$2.4B**	**$2.6B**	**$3.3B**	**$3.4B**	**$3.6B**
TOTALS	**$113.5B**	**$118.6B**	**$122.9B**	**$127.4B**	**$127.7B**

communications is how they will operate on a country to country basis, especially when they seek to sell services to end-users and whether local telecommunications service providers will insist on a share of the revenues. It was this critical factor that created the problem for Iridium and Globalstar in obtaining landing licenses in countries around the world. The requirement to operate through local telecommunications providers greatly inflated the costs of these satellite service providers when they sought to operate on a retail basis as opposed to a wholesale basis, which is most common for fixed satellite service providers who typically sell transponders or managed satellite services to local telecommunications providers.

Regulatory Oversight Concerns

And there are more than just market concerns related to all of these new LeoSat constellations. There are also serious concerns related to space traffic management and orbital space debris issues that are also worthy of serious policy analysis. Many now feel that new regulatory action is needed at the national and/or international level.

There are definitely increased policy concerns that come with the prospect of perhaps tens of thousands of new satellites being launched into low Earth orbit. How will these satellites be de-orbited at the end of life? What are the implications if a defunct satellite, like the defunct Russian satellite that crashed into the Iridium satellite in 2009, should recur? In such a case would it set off a cascade of collisions within these new satellite constellations? If all of the proposed satellites were actually launched this would increase the number of satellites in orbit by more than a factor of ten. All of the proposed systems have identified methods to control their own network and to de-orbit satellites and to avoid interference to the protected class of GEO communications satellites, but there is no defense against defunct, out of control satellites already in space and particularly concentrated in the polar regions where Sun-synchronous meteorological satellites are launched and where many defunct satellites now orbit.

Fig. 2.4 provides a graph that shows by type the growth of tracked satellites and orbital debris of significant size and the corresponding increases over time. This graph shows the two significant impulse increases in debris that occurred when the Chinese shot down one of their own defunct weather satellites in 2007 and then again in 2009 when a defunct Soviet weather satellite collided with a functioning Iridium satellite.

The current projection is that even without additional launches another collision that creates major new debris will occur on average every five to ten years. The European Space Agency using a computer-based simulation model has concluded that a collision will likely occur every five years, while NASA models project collisions somewhat less frequently. At the time these ESA estimates were first presented at an orbital debris conference in Frankfort, Germany, by Dr. Klinkrad, at the time Head of ESA's orbit space debris unit. He said: "The only way to keep this from happening is to go up there and remove them. The longer you wait, the more difficult and far more expensive it is going to be." [8].

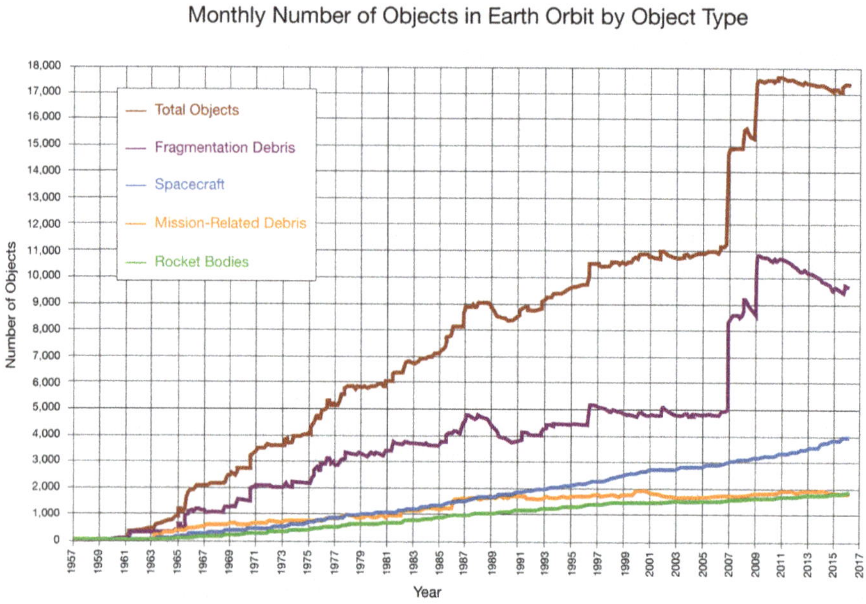

Fig. 2.4 Objects larger than 10 cm in diameter being tracked in Earth orbit. (Chart courtesy of NASA.)

Orbital space debris and space traffic management are issues that the U. N. Committee on the Peaceful Uses of Outer Space, and especially its Working Group on the Long Term Sustainability of Outer Space Activities (LTSOSA) are addressing. The book *Global Space Governance: An International Study*, by this author, was published in 2017. It includes the result of a truly international study that was conducted between 2014 and 2017 on a number of key space issues and recommended actions focused on these issues and possible actions that might be taken. Chapters 13 and 14 particularly addressed the topics of space traffic management, including not only for Earth orbit but also for near space (called the protozone), orbital space debris and on-orbit services as well as active debris removal [9].

This study recommended that the U. N. Committee on the Peaceful Uses of Outer Space (COPUOS), and the U. N. International Civil Aviation Organization (ICAO) seek to devise a framework where international guidelines for implementing space traffic management for both Earth orbit and the protozone might be undertaken. Such consultative processes need to be undertaken, in cooperation with their member states and interested bodies and organizations such as the InterAgency Space Debris Committee (IADC), Secure World Foundation, the International Association for the Advancement of Space Safety (IAASS) and the International Telecommunication Union (ITU). The key missing step in this process is the lack of clear-cut international agreement as to how to proceed. Perhaps there needs to be a new international treaty, an amendment to the Chicago Convention of 1944 or some other agreement, perhaps reached within the U. N. General Assembly on the recommendation of COPUOS [10].

What is clear is the need to create a new globally agreed framework whereby space-faring nations might agree to cooperative arrangements for space traffic management (both for Earth orbit and for the protozone) and for active space debris removal to be undertaken. This might also require new interpretations of the provisions of the Liability Convention and other international agreements as to who (i.e., nation states, private commercial organizations under licensing by nation states, or designated international entities) might undertake these activities.

The plans to increase operational satellites from the current 1,500 or so spacecraft to as many as 15,000 and the ever-increasing risk of orbital collision as well as many new possible activities in the protozone region, makes action in this area of even greater importance. Ever expanding interest in the protozone also creates concerns as well. These national and commercial interests include operation of spaceplanes taking suborbital flights, positive hypersonic transportation flights, high altitude platforms for communications, networking and remote sensing, high altitude launch of rocket launchers and spaceplanes, robotic transport flights above commercial airspace, and possible dark sky research platforms with electronic propulsion flights to orbit.

It would be most unfortunate if international agreement and positive proactive action is not taken soon within the international space community and well before a catastrophic accident or runaway space debris cascades as predicted by the so-called Kessler syndrome becomes a reality.

Conclusions

There is no area of space applications that is currently more dynamic, more churning with technological innovation, or larger in market size than that of satellite communications. Change is everywhere, but the outcome in both market direction and technological success is far from clear. There are innovations in Earth station design and new technologies and systems being rapidly developed to support large-scale constellations. There are new capabilities to launch telecommunications and networking satellites into orbit at lower cost. The advent of reusable rocket launchers is of particular note. There is great innovation that comes from additive manufacturing, 3D printing, large-scale manufacture and automated quality testing that is allowing the building of satellites, Earth stations and launchers faster, at lower cost, and hopefully with greater reliability.

What is clear is that there are new entries into all aspects of the space industry. Many of these new initiatives cluster around the space sector with the greatest revenues, potential profits, and perhaps greatest growth potential. Time will tell if the projected new markets to bring network connectivity to the underserved developing countries will pay off as many are anticipating.

The next five years will show whether the established satellite providers will adapt successfully and well to this new environment or whether the many new entrants will emerge as the new stars in the dynamic world of satellite communications applications.

References

1. The space report. The Space Foundation. https://www.thespacereport.org/ (2017). Accessed June 2018
2. Northern Sky Research market briefing on new satellite constellation. 7th Annual Space and Satellite Consortium, ReedSmith, Washington, D.C., 18 October 2018
3. Kymeta electronically steered antennas. https://www.kymetacorp.com/. Accessed 25 August 2018
4. Henry, C.: Phasor sets 2018 release for electronically steered antennas. Space News. https://spacenews.com/phasor-sets-2018-release-for-electronically-steered-antenna/ (4 Aug 2017)
5. DiNapoli, J., Baker, L.B.: Exclusive: SoftBank to let OneWeb-Intelsat merger collapse. Reurters. https://www.reuters.com/article/us-intelsat-m-a-oneweb-exclusive-idUSKBN18S3LP (31 May 2017)
6. LeoSat: a new type of satellite constellation. http://leosat.com/technology/ (Accessed 9 June 2018)
7. Satellite Industry Association: State of the industry report for 2017. Bryce Space and Technology. https://www.sia.org/wp-content/uploads/2017/07/SIA-SSIR-2017.pdf (June 2017)
8. ESA: Time to clear space junk from Earth's orbit, Reuters. https://www.voanews.com/a/european-space-agency-esa-space-junk-removal-earth-orbit-gps/1648848.html (25 April 2013)
9. Jakhu, R., Pelton, J.N.: Global Space Governance: An International Study. Springer, Basel (2017). Chapters 13 and 14
10. Convention on International Civil Aviation, 7 December 1944, 15 UNTS 295, Can TS 1944 No. 36 ICAO DOC 7300/9 (Chicago Convention)

3 Key Trends in Remote-Sensing Satellite Systems and Services

Introduction

Remote sensing is an extremely valuable tool that is vital to many industries, from mining to fishing and farming to mapping and surveying operations. Indeed the uses of Earth observation satellites spread across a wide spectra of other industries. The ways to use remote sensing just keeps expanding. Each year new applications are being developed in fields as diverse as real estate development, retail sales, urban planning, law enforcement, disaster relief and recovery, archaeology, energy exploration, and transportation. There is even an online listing of the top 100 uses that can now be made of remote-sensing satellites [1].

There are basically two types of tools used in Earth observation – passive and active. The most common type of space-based remote-sensing uses reflected sunlight and energy, or what is called passive sensing. The other approach uses active sensing systems. Active systems require both transmitted energy to be sent down and also requires a capability to receive the reflected signals that come back up to the satellite. The passive systems use optical sensors and photography, infrared sensors, charge-coupled devices and even radiometers for higher spectra. Active systems that require the transmission of energy, which is then reflected back to the satellite, include RADAR and LIDAR. All of these space-based remote-sensing satellite systems have been fine-tuned and improved to provide increased accuracy of observation. Over the course of a half century, remarkable progress has been made to improve the accuracy of remote sensing and the analytical capabilities that can be used to interpret this data.

Weather or meteorological satellites are a special form of remote-sensing or Earth-observation satellites. Since these satellites, which monitor Earth's weather and now also monitor climate change and solar weather, are so vital to modern civilization and are exclusively operated as governmental facilities, these satellites will be addressed later in a separate chapter.

Remote-sensing satellites have not only provided vital assistance to many commercial industries, but they have aided government in providing vital

J. N. Pelton, *Space 2.0*, Astronautical Engineering,
https://doi.org/10.1007/978-3-030-15281-9_3

services. These services range from law and regulatory enforcement, prosecuting crimes against humanity, environmental and pollution monitoring, to providing assistance in areas such as agriculture, forestry, mining, water management and fishing regulation. Of all these governmental functions perhaps the most important of all has been in the area of providing assistance with regard to disaster prediction, response and mitigation.

There are four primary ways of increasing the precision of Earth observation – or remote observation of other planets for that matter. These include: spatial (i.e., more pixels per area of measurement), spectral (i.e., the width or precision of the frequency spectrum bands where the observations are made), temporal (i.e., how frequently the observations are updated), and radiometric (precision of the data recorded in terms of available categories to collect data).

Radiometric precision started with just a 6-bit data recorder (equivalent to 64 bits of data). This then moved up to an 8-bit data recorder (equivalent to 256 bits), and today information is typically collected with a 10-bit data recorder (equivalent to 1,024 bits). Over time, all of these scales of precision have increased dramatically. There are civilian systems today that record pixels of data spatially equivalent to 0.35 meters and defense-based systems that are even more precise. Hyper-spectral systems can collect spectral data with great precision by recording data over as many as 100 different discrete spectrum bands, if not more. New constellations with hundreds of small satellites can update data with amazing frequency down to hours within a day. The most modern data recorders use 10-bit systems or even higher.

There are a number of obvious reasons that have given impetus to growth in the field of remote-sensing satellites; these have served to move it forward over the past 50 years. These reasons are combining to create not only more precise observations but innovation in how the various systems are applied to solving problems or assisting both government and industry to perform their various tasks. Although remote-sensing economic activities are much smaller than satellite communications in market size, significant growth is forecast for the future. One market study has concluded that the Global Remote Sensing Services Market accounted for $8.68 billion in 2017 but is projected to rise sharply in the next decade to reach $38 billion by 2026. This would represent a cumulative aggregated growth rate of nearly 18%. This study included not only satellite-based growth but also that of aeronautical sensing, drones and high-altitude platform systems [2].

The growth factors include: (a) the development of small satellite technology and the lower costs of manufacture, testing and launch of all types of remote-sensing satellites; (b) new digital processing and analysis techniques and especially the development of new hyper-spectral sensing capabilities; (c) spinoffs from defense-related surveillance; (d) new commercial business interests around the world; and (e) strong political motivation to have national remote-sensing satellite capabilities in space among a growing number of countries.

Reinventing the World of Remote Sensing

The world of remote sensing, just like the world of satellite communications, is being turned upside down. Disruptive technologies and new commercial applications for remote-sensing services are seemingly everywhere. This is true with regard to the small satellite innovations, new thinking with regard to launch services, new commercial applications, and in the context of the new entrepreneurial spirit born of the Space 2.0 revolution.

The Small Satellite Revolution in Remote Sensing

Many of the changes that are now occurring in the world of space applications have come from the world of cubesats. This has already been described as the coming together of the world of Silicon Valley with the world of aerospace. This Space 2.0 mentality is driven by the zeal for entrepreneurial innovation meshing in new and disruptive ways of doing things in the space industry around the world.

It turns out that many of the key components needed for smaller and more efficient remote-sensing satellites – namely digital processors, digital imaging devices, charged-coupled devices, radiometers, and digital sensors – are all possible to miniaturize. The young innovators who came out of Silicon Valley and who, in particular, founded Skybox and Planet Labs, asked why not invent a new way of doing things? They brashly thought that they could design and build small satellites that were more cost effective than the conventional remote-sensing satellites that had been growing in size and performance since the first systems such as Corona and Landsat were designed and built a half century ago. It turns out that they were right.

These new entrants thought their "good enough" technology, designed with miniaturized components and in some cases using off-the-shelf technology, could be used to build successful small satellites. Their small sats actually could provide commercial competition to the big and expensive remote-sensing satellites. Their innovative plan was to build satellites that could be ten to even a hundred times less costly than the much bigger remote-sensing systems that had evolved over a forty-year period.

Communications satellites need high gain and thus large antennas and more power, but remote-sensing satellites, with their "shrunk" electronic sensors and imaging device were a logical fit for creating a small sat fleet for remote imaging. Further, their data from imaging did not have to be continuously linked to Earth stations. Also, their constellations circling in low Earth orbits allowed their sensors to have higher spatial resolution. These were some of the technological reasons why small satellites for remote sensing, especially passive systems, have some advantages over telecommunications satellites.

Actually Skybox and Planet Labs were not the only innovators to led the way. The first champions of small satellites were the scientists, educators and engineers at the Surrey Space Centre at the University of Surrey just outside of London, England. The University of Surrey small satellites were known as UOS satellites. The spinout group, known as Surrey Satellite Technology Ltd (SSTL), was created in 1979. It started with designing very small satellites similar to the OSCAR satellites for store-and-forward messaging that were first used by amateur radio operators. SSTL, which is now majority owned by Airbus, then moved on to remote-sensing satellites. Their Disaster Monitoring Constellation (DMC) satellites (in 2002) and their UoS-12 remote-sensing satellite (2006-2008) led to a series of contracts or technical cooperation agreements to design, build or support the building of remote-sensing satellites [3]. Technical support from the Surrey Space Centre and SSTL in the design and manufacture of small sats allowed countries such as Nigeria (NigeriaSat-2) [4], Korea (Arirang-1 in 1999) [5], China, and other countries to deploy relatively small and cost-effective small remote-sensing satellite in the 1999-2010 timeframe.

Progress in the small sat field for remote sensing has recently moved ahead very swiftly. In one of the latest examples a new commercial system was developed for precise imaging. This Chinese Jilin satellite system consists of a three-satellite constellation. In this case, SSTL provided the spacecraft bus design for the three-satellite constellation while China designed the imaging payload. These three satellites were launched from India by ISRO in 2015. Although these were small sats with only 450 kilograms in mass, each nevertheless have an amazingly high spatial resolution of 1 meter [6].

The Chinese Jilin-1 satellite, which was based on an SSTL spacecraft design but contains a Chinese imaging payload, is the first commercial Chinese remote sensing satellite system now fully operational [7]. The first three satellites in the satellite constellation, now operated by the Chang Guang Satellite Technology Company, will be joined by a fourth higher performance remote-sensing satellite designed and built entirely within China. The remarkably high spatial resolution of images taken from this constellation at 1-meter resolution shows amazing detail. (See Fig. 3.1 that shows a precise image

Fig. 3.1 Image of the Ferarri Exhibit in UAE from China's Jilin-1 commercial imaging satellite. (Illustration courtesy of NASA Spaceflight, Global Commons)

taken from the Jilin-1 satellite over the Ferrari Exhibit in Dubai, United Arab Emirates.)

In addition to the Jilin commerical satellite, there is an even more ambitious Chinese commercial project now underway. Satellites for both video and hyperspectral remote-sensing purposes are currently being deployed via Long March 11 solid rocket launchers. On board the most recent April 2018 launch were one high-resolution video coverage satellite and four hyperspectral Earth-observation satellites. These satellites were for the Zhuhai Orbita commercial constellation and were designated as OHS-01, 02, 03 and 04 (hyperspectral) and OVS-2 (video).

This system is being deployed by Zhuhai Orbita Aerospace Science and Technology Co. Ltd. This Chinese remote-sensing company is based in the city of Shenzhen. The OVS-1A and OVS-1B satellites were launched in June 2017, and on April 26, 2018, five more satellites were launched. One of these was the improved OVS-2 video satellite that has a spatial resolution of 90 centimeters. In addition four commercial hyperspectral satellites were also launched with a resolution of 10 meters. Eventually, a full constellation of 34 satellites orbiting at 500 km altitude is planned to be launched by the Zhuhai Orbita company. This combined constellation of high-resolution video satellites and lower resolution but hyperspectral scanning satellites is designed to provide geographic, environmental and geologic monitoring as well as coverage related to marine and urban planning use. Small sat efficiency and economies has seemingly served to accelerate commercial remote-sensing satellite launches around the world in the last few years (http://spacenews.com/china-launches-five-commercial-remote-sensing-satellites-via-long-march-11/).

The advent of small satellites and original thinking that comes from the world of NewSpace keeps spawning completely new ideas. One of the latest space innovations is the Hawkeye 360 project. This new remote-sensing constellation plans to deploy 30 small sats that consist of ten groups of three satellites that will be monitoring radio frequency usage on a global basis. This satellite constellation will collect data on the usage of mobile cellphones and satellite usage that will provide worldwide updates every 30 to 45 minutes. Such monitoring can track usage and movements of ships and boats, illegal fishing, location of jamming systems and hundreds of other new applications. The U. S.-based startup is a subsidiary of Allied Minds, which is a 'venture creation' firm and is also teamed with Lockheed Martin [8].

For many decades the lead in remote-sensing satellites remained with the United States and other countries of the OECD and particularly with NASA, NOAA, and ESA, but the small satellite revolution has served to change the world of remote sensing. Many other countries are designing, manufacturing, and deploying or arranging for the launch of their own remote-sensing satellites. Some new satellite systems, such as Theia, will be offering remote sensing, communications services and data analytics services to create what might be considered a whole new category of satellite services [9]. The deployment of so many of these new satellite constellations, particularly in Sun-synchronous polar and LEO orbits, contributes to concern about the continuing increase in orbital space debris.

The number of remote-sensing satellites continues to increase. In the latest SIA State of the Industry Report, remote-sensing satellites represent some 15% of all operational satellites. A decade ago that number was well below 10%. With the growth of commercial constellations comprised of small satellites, the number of commercial communications satellites and remote-sensing satellites will only continue to grow. The Indian Polar Satellite launch vehicle that placed 88 3-unit cube satellites for the Planet Corporation, along with other cubesats for other customers and some larger satellites all into low Earth orbit with a single launch, perhaps portends a future with an dramatic increase of small sats in Earth orbit [10]. (See Fig. 3.2.)

One of the more interesting international developments has been the recent agreement reached at a meeting in Haikou, China, in July 2017 by the five BRICS countries – Brazil, Russia, India, China and South Africa. These countries will develop a Brics Remote Sensing Satellite Constellation under the first substantive BRICS cooperation agreement related to space research (http://www.engineeringnews.co.za/article/brics-bloc-agree-remote-sensing-space-constellation-project-2017-07-04/rep_id:4136). (See Fig. 3.3.)

The Brazilian CBERS-4 satellite, which is pictured in Fig. 3.2, is under construction as a joint project between Brazil and the Chinese National Space Administration (CNSA) and is to be deployed into the virtual constellation. The CEO of the South Africa National Space Agency, Dr. Val Munsami, made the following statement at the end of the sessions held in China about the new BRICS initiative in space cooperation: "We remain committed to ensuring the integration of African space-based knowledge and technology in improving the lives of fellow Africans and welcome such esteemed partners in achieving this important objective" [11].

In Phase 1 of the project, the constellation would be a virtual network with the five countries creating a remote-sensing data sharing system. This would involve a network for sharing data from the countries' existing Earth observation (EO) satellites. Phase 2, as now planned, would involve the creation of a new EO satellite constellation. Of the five countries only South Africa does not have a full-scale remote-sensing satellite. It does have plans to deploy, either in 2019 or 2020, such a full-sized spacecraft known as EOSat-1 and which is currently under development.

Currently there are many dozens of remote-sensing satellite networks in operation on behalf of governments and private companies. The idea of consortia and international partnerships can aid not only international cooperation but also curtail the needless proliferation of satellite remote-sensing satellites in Earth orbit and the increasing problem of orbit space debris plus the higher risk of orbital collisions.

Developments in Digital Processing and Analysis of Remote Sensing Satellite Data

There are several counter trends in the remote-sensing satellite industry when it comes to the analysis and distribution of remote-sensing data. There are quite

Fig. 3.2 The Indian Polar Satellite launch vehicle that sent up a record number of 104 cube satellites in Feb. 2017. (Illustration courtesy of the Indian Space Research Organization.)

different needs with regard to the processing of remote-sensing satellite data when it comes to the requirements of various users in government, of different types and sizes of industry, and individual consumers. We are seeing increases in the number of satellites deployed. We are seeing even small sats with improved spatial resolution. Finally we are seeing various types of hyper-spectral satellites that are monitoring as many as a hundred small segments or bands of spectra. This can result in more timely updates of data. It can require higher

Fig. 3.3 The joint Brazilian/Chinese CBERS-4 satellite will form part of the planned BRICS virtual EO constellation in Phase 1. (Illustration courtesy of INPE of Brazil.)

performance data recorders, and faster downloads of data plus the need to cope more efficiently with the streams of encoded data. Overall these innovations has created a Niagara of data streaming from space that is difficult to process efficiently.

The huge spikes in data create a giant challenge to timely and thorough data analysis. Two key possibilities are on-board processing as well as AI algoritms that allow the 'automated' analysis of data based on expert systems and even more advanced AI 'reasoned' analysis. The NASA remote-sensing project named "Mission to Planet Earth" that began in the 1990s had to be reduced in scope just because the processing and analysis involved so many petabytes of data that it was impossible to cope with the downloading and analysis requirement. This is just one of the many instances where the hardware design for a new scientific mission and the supporting software can find themselves out of synch [12].

This push toward automated analysis and pre-processing has set off a controversial debate as to how much "ground-truthing" of remote-sensing data and expert analysis is needed to attain scientifically accurate results versus quickly providing analysis in near real-time to end users. There are fears that such end users, not well versed in data analysis, may be led into false conclusions from quickly downloaded pre-processed data that might tend to produce spurious results and wrong conclusions about such things as soil conditions, tree blight, oceanic pollution, and so on.

As the amount of data being produced by remote-sensing satellites rises from terabytes to petabytes to exabytes and beyond it will be increasingly likely that AI algorithms and other forms of pre-processing will be needed with the giant streams of data traffic. Thus in this area, as in many other aspects of a super-automated future, for great care to be given to human-machine interfaces (HMI). Such precautions can hopefully prevent major errors from being made.

One of the key software concerns is the now critical need to make sure that cyber-security systems are carefully put in place. Recent studies have shown how accessing aircraft systems could provid a means to cause an airplane crash. It is, of course, equally true that such a hacking of spacecraft avionics could be used to weaponize a spacecraft and to cause it to de-orbit or possibly crash into another space system [13].

Spinoffs from Defense-Related Surveillance

The first remote-sensing or Earth-observation satellite systems were designed and deployed for military and defense-related surveillance purposes. The very secret U. S. Corona spy satellite program was operational for many years before it was ultimately declassified. What were once highly classified spy satellite capabilities, however, are now routinely provided on commercial remote-sensing satellites such as Ikonos. Even the Chinese commercial remote-sensing capabilities provided on small satellites, such as the Jilian satellites at 1-meter resolution, were once considered to be spy satellite capabilities. The United States has reserved the right for ‘shutter control’ on commercial satellite systems with high spatial resolution that would serve to blank out areas where military hostilities might be taking place. This type of shutter control, however, tends to become a moot point if there are many different countries operating governmental or commercial satellite systems with equivalent spatial resolution.

It is also true that some countries seek to hide sensitive military installations or facilities by making them deliberately fuzzy on remote-sensing satellite images. This, however, becomes somewhat self-defeating in that hostile forces only have to look to the fuzzy regions on remote-sensing images to identify where these facililties are located.

Article 4 of the U. N. Outer Space Treaty of 1967 states in part: “States Parties to the Treaty undertake not to place in orbit around the Earth any objects carrying nuclear weapons or any other kinds of weapons of mass destruction, install such weapons on celestial bodies, or station such weapons in outer space in any other manner.

> The moon and other celestial bodies shall be used by all States Parties to the Treaty exclusively for peaceful purposes. The establishment of military bases, installations and fortifications, the testing of any type of weapons and the conduct of military manoeuvres on celestial bodies shall be forbidden.....

This treaty does not define what is a space weapon *per se*. It also does not explicitly prohibit the placing of satellites into Earth orbit that might be used to support military communications, operations, targeting, surveillance or reconnaisance. Today, the GPS and other GNSS networks can be used for the targeting of bombs, while other remote-sensing and Earth-observation satellites can be used for surveillance and reconaissance, and there are many satellites whose function is to support military communications and tactical operations. There are many remote-sensing satellites – both commercial and governmental – with spatial resolutions in the range of 0.35 m to 1.00 m that could be used for either offensive

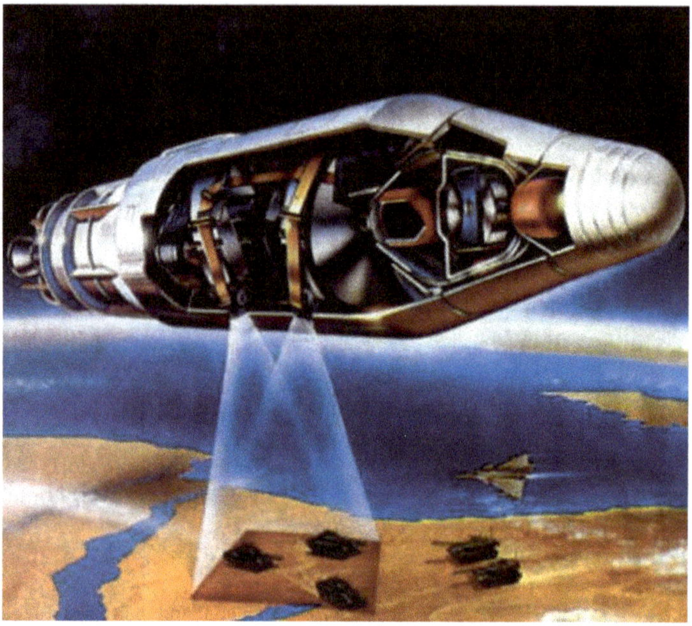

Fig. 3.4 The Corona program for defense spying led the way for civilian remote sensing in the years that followed. (Graphic courtesy of Yahoo History Chronicles.)

or defensive action by armed forces, and most are not subject to shutter control in the case of war. When the 50th anniversary of the U. N. Outer Space Treaty was celebrated in 2017, no international law expert foresaw the opportunity or even any prospect that new treaties to define space weapons or to ban their use could be agreed in today's political climate. The recent proposals that have come from the U. S. White House in mid March 2018 to create a U. S. Space Force has clouded the waters even further. Some believe that such proposals further jeopardize the ability of the nations of the world to respect the concept of space being a weapons-free zone [14].

Today, the eyes in the sky only continue to become more clear-sighted. If any nation seeks to hide strategic facilities or equipment, they tend to move it underground and thus out of sight of surveillance or remote-sensing satellites.

New Commercial Business Interests Around the World

The latest breakthroughs are now even allowing the deployment of small satellites that are capable of hyperspectral sensing in many dozens to over one hundred different bands and quite reduced-sized satellites able to provide video color imaging at spatial resolutions under 1 meter per pixel. These developments are changing the economics, business plans and applications of the world of remote sensing. The costs of these operations, that have come down by a factor of ten to even a hundred times, means that these services can be provided to a much broader range of users.

Farmers are increasingly seeing that the marginal cost of obtaining hyperspectral information for their farmlands makes a good investment to see how they can change their use of fertilizers and watering systems to increase productivity by 10% to even 25%. Entirely new applications are emerging. The rapid coverage provided by large-scale constellations such as Planet can allow retailing organizations, gambling casinos, and others to now track the parking in not only their own parking lots but also those of their competitors.

The challenge of the operators of today's remote-sensing satellite systems is not how to improve the imaging technology of their space systems but to find new commercial users for their increasingly detailed and useful data that comes down at a faster and faster rate. The key is in the analysis, distribution and sales of their space-based observations.

An entirely different concern is that as remote-sensing satellite services become more and more commercialized by entrepreneurial companies, the key role that remote satellites can play with regard to detecting disasters and providing assistance to disaster relief might become more limited. The International Charter on Space and Major Disasters was signed in the year 2000 and now has16 member agencies. These space agencies are committed to contributing free satellite, data processing and data distribution assets when this charter is formally invoked. It aims to provide a unified system of space data acquisition and delivery to those affected by natural or manmade disasters through authorized users.

The International Charter on Space and Major Disasters is a worldwide collaboration among space agencies, through which satellite-derived information and analysis products are made available to support disaster response efforts. The charter has been operational since November 2000, and currently the following global space agencies participate: ESA, CNES, CSA, NOAA, CONAE, ISRO, JAXA, USGS, UKSA & DMCii, CNSA, DLR, KARI, INPE, EUMETSAT, and ROSCOSMOS. This charter process involves only national and regional space agencies and official governmental agencies involved with remote-sensing data [15].

In the future it might certainly make sense to see if there were a process whereby a number of the commercial remote-sensing concerns that may in many instances have more rapid temporal updates of affected disaster areas might be included within the charter to provide vital data and analysis on an emergency basis to aid recovery efforts in a more timely way.

The Proliferation of Remote-Sensing Satellite Systems

The spread of remote-sensing satellite systems operated by national governments and a rapidly growing number of commercial satellite systems is one of the major stories in the rapid growth and expansion of the satellite world and the burgeoning Space 2.0 story.

For many years there were only a few spacefaring nations, i.e., less than ten, and this meant that there were a limited number of remote-sensing satellite networks, even with the inclusion of meteorological satellites. The world of small sats, which that has seen the rapid growth of RS satellites with hundreds of 3-unit cube satellite networks in operation and many other small satellites now operating as commercial operations, has dramatically altered the "spacescape" in the past decade. Today there are some that are questioning the need for quite so many satellite networks both for remote sensing and for space navigation and timing. Such networks are still expensive to design, build, operate, and even to de-orbit. There is particular concern about the deployment of so many satellites in low Earth orbit, and particularly in low Earth orbit Sun-synchronous polar orbit in terms of proliferating space junk and the rising threat of increasing orbital space debris.

The motivation behind the deployment of these networks continues to be various in number and not the same from country to country or from company to company. Many countries feel they need such types of space systems from the perspective of national defense, and particularly feel they need their own space resources rather than be dependent on other countries' resources. This is a motivation that is probably even stronger with regard to space-based precise navigation and timing satellites. Some countries or commercial companies feel they have special sensors or technical capabilities that are not available on other systems. Yet others believe that, with the latest small sat technology and automated testing capabilities and new, more cost-effective launch systems that can deploy new space systems at a cost that is perhaps as much as a hundred times less than was the case before, they can afford to do this themselves where once this was not the case. The motivation may be particularly strong for countries or commercial entities that see such satellite deployments as "flying their flag" in space and thus be able to display a special sense of accomplishment. (Fig. 3.5 shows the proliferation of remote-sensing satellites in recent years, and this chart is far from complete, with hundreds of new remote-sensing satellites launched in just the last two years.)

The greatest concern is with regard to the increasing deployment of both communications satellites and remote-sensing satellite constellations. This rapid increase of satellites in low Earth orbit exacerbates the problem of orbital space debris. With the current lack of clearcut and international agreements for space traffic management these problems will likely only become worse. Efforts to coordinate remote-sensing activities among nations and share data more freely are positive steps that can be taken.

In this regard one of the key developments is the global coordinative processes that are now in place through the International Committee on Global Navigation Satellilte Systems (GNSS) that is known as the ICG. The stated purposes of the ICG are to: "encourage coordination among providers of global navigation satellite systems (GNSS), regional systems….to ensure greater compatibility, interoperability, and transparency, and to promote the introduction and utilization of these services and ….encouraging coordination and serving as a focal point for information exchange." [17] The ICG came into existence through the good offices of the U. N. Office of Outer Space Affairs and has been in existence since 2005. If there are more processes for information exchange and more transparency and interoperability, this can perhaps help to minimize additional redundancy among remote-sensing satellite launches [18].

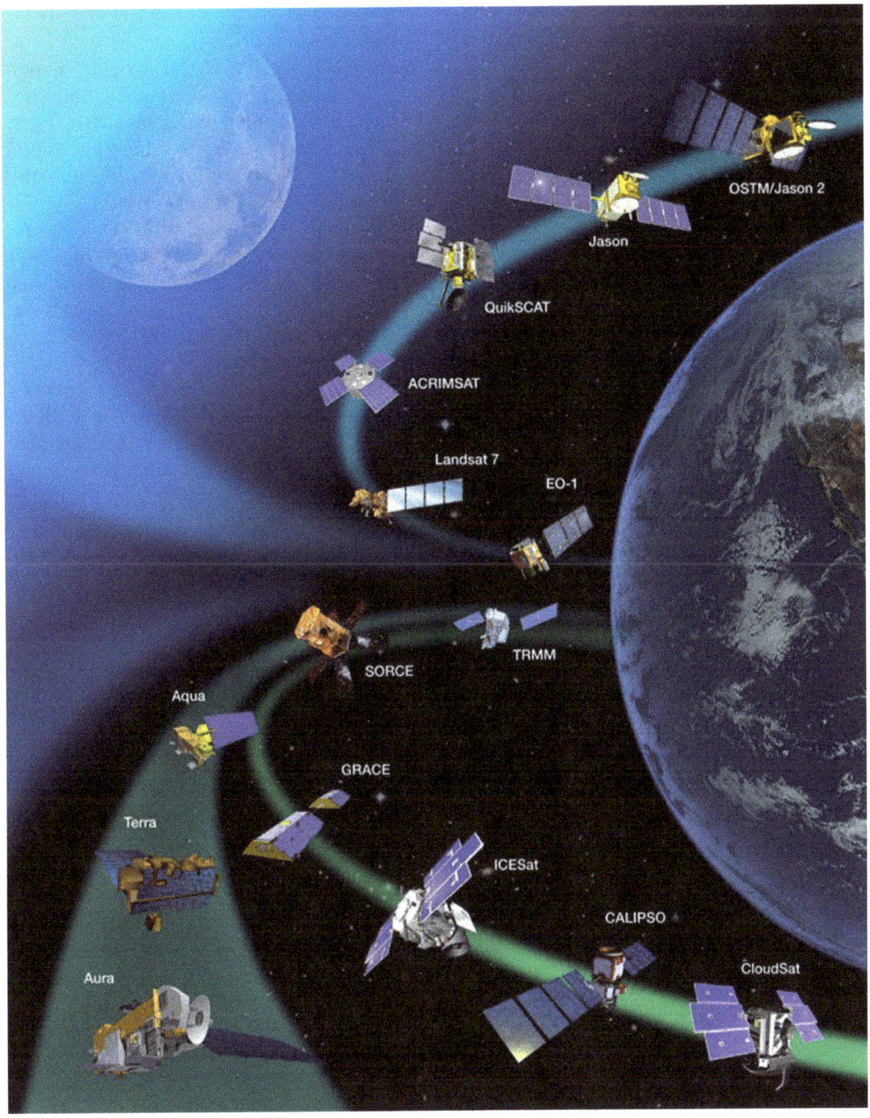

Fig. 3.5 The proliferation of governmental and commercial rs satellites globally. (Illustration courtesy of SERC, Carleton.edu.) [16]

Conclusions

The current rapid reinvention of the space industry and the growing success of Space 2.0 enterprises are particularly evident in the area of remote-sensing satellites. The strong influence that Silicon Valley has had on the space industry is particularly strong in the remote-sensing sector. The Skybox and Planet Labs initiatives were born of Silicon Valley ventures. Google purchased Skybox, renamed it Terra Bella and then facilitated the merging of the assets of the Skybox satellite system with Planet Labs to become the new Planet system. Hawkeye 360 is in the process of developing an entirely new form of satellite monitoring via small sats to track global use of mobile radio frequencies.

There are enormous amounts of data flowing from an unprecedented number of remote-sensing satellites in orbit, and the applications keep growing. The appendix to this chapter provides a guide to some of the databases that are now available free of charge that can be easily accessed by the Internet.

In addition to important new and growing industrial applications, governments are using RS satellite data for regulatory oversight, law enforcement, support to first responders, fire and environmental concerns and more effective response to disasters – both manmade and natural catastrophes. The challenge is to cope with the mounting problem of orbital space debris and in this context to develop a meaningful system for space traffic management that can minimize the risk and perhaps facilitate active debris removal.

See Appendix A at the end of this book for an addendum to this chapter.

References

1. 100 Earth shattering remote sensing applications & uses. GIS Geography. https://gisgeography.com/100-earth-remote-sensing-applications-uses/ (17 Feb 2018)
2. Remote sensing services market by platform (satellites, UAVs, manned aircraft, and ground), end user (defense and commercial). https://www.researchandmarkets.com/reports/4419378/remote-sensing-services-market-by-platform#relb0-4541418 (Oct 2017)
3. Surrey satellite technology. https://en.wikipedia.org/wiki/Surrey_Satellite_Technology. Accessed 14 June 2018
4. NigeriaSat-2. eoPortal Directory. https://directory.eoportal.org/web/eoportal/satellite-missions/n/nigeriasat-2
5. Arirang-1 Satellite. https://en.wikipedia.org/wiki/Arirang-1. Accessed 14 June 2018
6. Sino-UK remote sensing satellite constellation launched, July 12, 2015. http://www.china.org.cn/business/2015-07/12/content_36040295.htm
7. Jones, A.: Jilin-1: China's first commercial remote sensing satellites aim to fill the void. GB Times. https://gbtimes.com/jilin-1-chinas-first-commercial-remote-sensing-satellites-aim-fill-void (12 May 2016)
8. HawkEye 360: meet the company ready to take its tech into orbit. Biz Journal. https://www.bizjournals.com/washington/news/2017/09/28/hawkeye-360-meet-the-company-ready-to-take-its.html (9 Sept 2017)
9. Theia Satellite Network: licensing.fcc.gov/myibfs/download.do?attachment_key=1158366. Accesses 23 Oct 2018
10. Mathewson, S.: India launches record-breaking 104 satellites on single rocket. Space.com. https://www.space.com/35709-india-rocket-launches-record-104-satellites.html (15 Feb 2017)
11. Campell, K.: BRICS bloc agree to remote sensing space constellation project. http://www.engineeringnews.co.za/article/brics-bloc-agree-remote-sensing-space-constellation-project-2017-07-04/rep_id:4136 (4 July 2017)
12. NASA's mission to planet Earth. https://www.hq.nasa.gov/office/nsp/mtpe.htm. Accessed 14 June 2018
13. Zetter, K.: Is it possible for passengers to hack a commercial airliner? Wired Magazine. https://www.wired.com/2015/05/possible-passengers-hack-commercial-aircraft/ (26 May 2015)
14. Borenstein, S., Ken Thomas Associated Press. Star wars? President Trump proposes military space force. Military Times. https://www.militarytimes.com/news/pentagon-congress/2018/03/14/star-wars-

president-trump-proposes-military-space-force/ (14 March 2018)

15. International Charter on Space and Major Disasters: http://www.un-spider.org/space-application/emergency-mechanisms/international-charter-space-and-major-disasters. Accessed 16 June 2014
16. University College of London: www.ucl.ac.uk
17. International Committee on Global Navigation Satellite Systems (ICG): http://www.unoosa.org/oosa/en/ourwork/icg/icg.html
18. Camacho-Lara, S., Pelton, J.N.: International Committee on GNSS. In: Pelton, J.N., Madry, S., Camacho-Lara, S. (eds.) Handbook on Satellite Applications, 2nd edn, pp. 765–780. Springer, Basel (2017)

4 The Growth and Expansion of Precise Navigation and Timing

Introduction

There are actually many terms that are used to characterize Global Navigation Satellite Services (GNSS) around the world. Some of these terms include: Precision Navigation and Timing (PNT), Sat Nav, and GPS (for Global Positioning Satellites). There are also a growing number of systems that provide this type of service. These include the Glosnass system operated by Russia, the Chinese Beidou and Compass systems, the Japanese Quasi-Zenith system, the Indian Regional Navigation Satellite system, and the European Galileo system. There are possibly others that might be initiated in the future, such as system deployed by the Republic of Korea and one operated by the United Kingdom if it is not able to negotiate continued participation in the Galileo system due to Brexit [1].

How GNSS Satellites Establish a User's Exact Location

A GNSS satellite is constantly sending out a timed signal, and as the signal radiates through space it creates an ever-expanding sphere. If at least three spheres from three different satellites intersect at the same instant in time it means it can only be one precise location above Earth's surface. The other possible intersection would be below Earth's surface. If four spheres from four satellites intersect at the exact same instant in time then there is only one possible location. If five spheres from five different locations intersect then there is even more precision as to the location. In short, the more signals from the overhead satellites that intersect at the same exact moment in time the better the accuracy of the location. Figure 4.1 illustrates the point – literally.

J. N. Pelton, *Space 2.0*, Astronautical Engineering,
https://doi.org/10.1007/978-3-030-15281-9_4

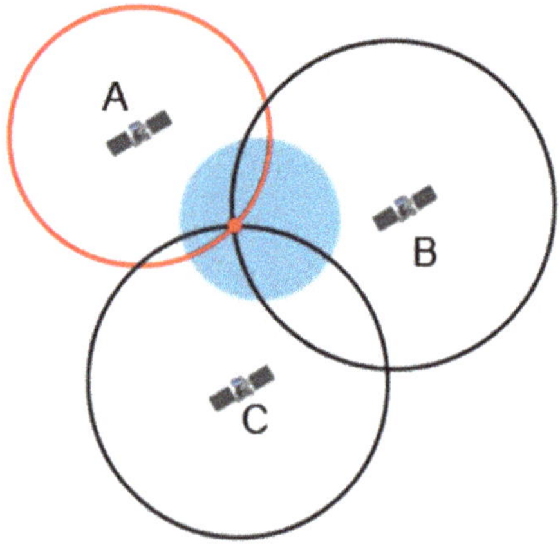

Fig. 4.1 The intersecting circles represent the radiated spheres from GNSS satellites

Innovations in GNSS Systems

These various navigation systems vary in their coverage and capabilities. All depend on the precision of timing provided by on-board atomic clocks. All but the Japanese system operate in high medium Earth orbits of about 12 hours in duration.

It is also possible to supplement the accuracy of these systems via land-based equipment to meet specialized needs. In the United States, the Federal Aviation Administration (FAA) has been responsible for the development of a satellite-based augmentation system (SBAS) known as the Wide Area Augmentation Service (WAAS). WAAS relies on ground stations located very precisely on very accurately surveyed locations distributed around the United States. It then constantly beams out locations via two different geosynchronous satellites, which are located at 107 degrees West and 133 degrees West.

The purpose of WAAS is to allow the combined GPS and WAAS navigation systems to assist with the takeoff, landing and all phases of aircraft operations in the United States. The stated goal of the WAAS operation is to improve the "accuracy, integrity and availability" of the GPS system as an aid to aviation and aviation safety [2].

There is also what is known as the Continuously Operating Reference Stations (CORS), which is run by the National Geodetic Survey (NGS). This unit is a part of the U. S. National Oceanic and Atmospheric Administration. CORS provides the NAVSTAR GPS network data consisting of carrier phase and code range measurements in support of three-dimensional positioning, meteorology, space weather, and geophysical applications throughout the United States, its territories, and also several other countries as well.

The CORS network is relied on by many surveyors and Global Information System (GIS) users, as well as a number of engineers, scientists and the public at large who are dependent on highly accurate positioning data, particularly those requiring three-dimensional locational data with exact precision within a very few centimeters – both horizontally and vertically.

The CORS network is a cooperative endeavor. Thus it involves many partners that include a number of governmental agencies plus academic, private and commercial participants. With the CORS network, the sites are independently owned and operated. There are over 2,000 stations in the network and over 200 different partner organizations [3].

There have been various efforts around the world to improve the accuracy of NGSS networks around the world via such means as ever more accurate atomic clocks and ground-based systems where exact locations are known, such as the WAAS and the CORS network. Improvements and augmentation systems that have been developed by other countries within their GNSS systems can be found in the article written by Dr. Sergio Camacho-Lara and referenced from the Second Edition of the *Handbook of Satellite Applications* by this author. Some of these systems are seeking to achieve even millimeter precision to support geodetic scientific purposes and other applications [4].

There have also been efforts to improve the resilience of GNSS systems and ensure their continuous availability and to protect them against possible cyber-attacks and hacking of their operating control systems.

One of the top concerns has been with an ongoing effort to upgrade the performance of GNSS satellites to make them less vulnerable to jamming and to guard against any hostile attacks that might be undertaken by a hostile government, a terrorist organization or even some form of criminal attack by hacker adversaries.

The velocity of the various GNSS satellites as they move in orbit relative to the speed of light is sufficient that the effects of Einstein's relativity of motion must be taken into account in order to achieve the proper calculation of a specific location with exact precision. Some of the most advanced systems now in operation include the Russian Glonass M series, the U. S. Navstar GPS Block IIIA satellites and the Chinese Beidou 3 satellites. (These three types of GNSS satellites are shown respectively in Figs. 4.2, 4.3 and 4.4.)

The specifications for the Chinese Beidou 3 satellites, for instance, include ultra-exact atomic clocks that are specified to have an accuracy of only being off by one second in every 300 years. Such accuracy, when coupled with high-performance ground systems, could produce location accuracies to within millimeters on Earth's surface [5].

If there is interest in the technical details about GNSS orbits, reference systems, carrier frequencies, time kinematic systems, GNSS Augmentation systems or other technical details, it recommended that one consults the article by Rogerio Enriquez-Caldera in the *Handbook of Satellite Applications* (Second Edition) [6].

In recent years the greatest areas of advancement have not come in the form of important technical upgrades to the satellites, No, the main advances are coming with the many important applications and new uses that continue to be developed for precise navigation and timing satellites. Today GNSS satellites provide vital services for the routing of ships at sea, cars, trucks, buses and other vehicles. These satellites assist with the takeoff, landings and routing of aircraft. They also provide vital timing services for countries around the world. The operational uses of these highly precise satellites also include time-stamping as

Fig. 4.2 The Glonass M Series as deployed by the Russian Federation. (Image courtesy of the Russian Academy of Science.)

part of the security system used for banking transactions and other secure financial transactions.

A market analysis of GNSS applications came up with the following figures. Location-based services represented 53% of the market, road-based services represented 38%. Other applications, such as surveying and mapmaking, rail, agriculture, fishing, maritime, timing synchronization, and security and aviation filled out the remaining 9%. The percentage of use, however, is not necessarily an accurate indication of the importance of the usage. The usage for aeronautical and time synchronization purposes only represents a little over 1% of the total, but these applications are vital to aircraft safety and operation of the Internet [7].

And going forward, opportunities remain even more diverse. GNSS systems will for instance not only be for aircraft routing and landing but will also be key to the operation of autonomous vehicles, autonomous freight operations, and even robotic freighters in the skies, flying above commercial airspace. This will not only potentially make automobile and trucking vehicular operations much safer, but all forms of freight operations on land and sea much more cost effective. There are many automated operations tied to GNSS networks that will likely allow significant productivity increases. Yet all of these vital operations come with additional concerns as to the reliable and continuous operation of these satellites. Thus there are significant privacy,

Fig. 4.3 The GPS Block IIIA Satellite Series. (Image courtesy of Lockheed Martin.)

cyber-security and vital infrastructure concerns attached to the safe and secure operation of these networks.

Key Issues and Concerns

The space-based precision navigation and timing services are now only second to satellite communications services in terms of market size. There are now billions of PNT units in operation around the world. Every smart phone in the world has a GPS unit installed inside, and thus there are truly billions of them operational worldwide. Just the manufacture and sales of GNSS units globally represents many billions of dollars in sales. The Satellite Industry Association's State of the Industry Reports for 2015 and 2016 put the totals for GNSS tracking units to be in excess of $30 billion per annum [8].

The applications have gone well beyond navigation, and the uses now include tracking of stolen vehicles and goods, security for tracking financial operations, support for detection

Fig. 4.4 Beidou 3 GNSS satellite with super accurate atomic clock. (Graphic courtesy of the China Academy of Sciences.)

systems for tsunamis, geodesic and seismic monitoring, and tracking of operational spacecraft and high altitude platform systems. In the future there might even be tracking of items such as orbital space debris or mobile robotic devices.

In the age of the Internet of Things, RFID systems, GNSS-based tracking and time-related services will be raised to a whole new level, with automated computer-based tracking of essentially anything that moves. The opportunities for GNSS-related services remain almost endless. Security systems of all different types can require someone to be precisely personally identified and also to be at an exact location to allow execution of various transactions. Commercial operations involved in transportation, shipping, retailing, farming, fishing and more can keep track of all of their assets in real time. Governments can use these systems for law enforcement, drug trafficking, pollution and environment policing and for a wide variety of national defense and military activities from surveillance to missile targeting to drone operations.

The concern is not whether we will find hundreds if not thousands of applications for GNSS networks. No, the concern is what is the backup plan if a vital GNSS system should fail or even be temporarily interrupted via jamming or other difficulty.

Should there be an interruption in the service of these vital networks, the security of nations and vital economic systems could be put at risk. Just one example of this criticality is the fact that for most countries of the world the synchronization of the global Internet relies on GPS, and thus if this system were to fail then the global Internet could be put at risk.

These new applications have significant opportunities for new commercial products and services, but there must also be serious concerns for the various issues that can arise.

Personal Surveillance, Privacy and Freedom

Today GPS-enabled cell phones not only allow efficient navigation from one location to another but it also allows precise tracking of where a person is located. Today in China, residents are assigned a Social Index Score that combines information about not only financial transactions and the prompt payment of bills but also data about crime convictions, traffic infractions, who one's friends are and political activities. This combined score indicates not only one's financial and legal status but also an overall assessment of whether one is a 'good citizen' or of questionable character. Instead of pressure to avoid 'bad' behavior, this scoring system creates a series of incentives to engage in "good behavior with good friends." [9]

And this is not a phenomenon that is unique to China. In the United States there is not only the FICO credit score, but a number of secret scores are also available. There is the Northpointe's COMPAS scoring system that is used to predict the recidivism rate for those going to jail and to assist in setting bail. In addition one can be judged by the Acxion consumer score. There is the Johns Hopkins Frailty Score, the SMR Research Charitable Donor Score, and the HiQ Labs Keeper Score that tracks the likelihood that employees might be recruited away by competitors. And this is far from a complete list. In the age of the Internet and GNSS systems it is possible to use computers to cross reference many different sources to track where one goes, what one does, how one spends money, and who they associate with. And this tracking capability will only increase with the spread of hundreds of billions of Internet of Things enabled units [10].

The growth of GNSS network usage is but one aspect of the digital privacy issue, but anyone who is carrying a smart phone with a GPS chip should be on notice that his or her whereabouts is constantly able to be tracked. In older movies and television shows the detectives were always asking for the whereabouts of suspects and checking out alibis. In today's world, one only has to have the smart phone's GNSS chip monitored. This should give pause to anyone who loans their cell phone to someone else or agrees to change the Subscriber Identity Module from one phone to another.

Cyber-Security and Hacking of Computerized Systems with GNSS-Enabled Devices

The other large and potentially even more significant concern about GNSS-enabled tracking systems is the possibility that they might be hacked by techno-terrorists, with catastrophic results. There are concerns that GNSS location results might be hacked in order to give airline pilots and autopilot systems false readings that could lead to fatal airline crashes. One of the James Bond movies *Tomorrow Never Dies* has the fantastical plot line with Johnathan Pryce playing a megalomaniac news executive, serving to distort GNSS readings so as to start a war between China and the West. This rather absurd plot line seems to suggest that he could make enormous amounts of money by reporting on the war that he rather singlehandedly has served to engineer.

Cyber-security is a difficult subject because there are so many potential areas of vulnerability. There are many types of malware in the form of viruses, Trojans, spyware and ransomware such as Wannacry. There are also so many possible modes of attacks through so many different types of devices such desktop computers, laptops, notebooks, cell phones and even data stored on the cloud, that users forget that there are even more potential areas of cyber-attack. The truth is that there are always more and new ways to possibly be attacked. There are now true concerns about Internet of Things-enabled devices from smart appliances such as refrigerators to washing machines to even more unlikely areas of vulnerability such as 'smart' doorbells, aquariums, or even baby monitors. There are certainly ways whereby GNSS systems can be jammed or potentially manipulated. And no one should forget that there is always some form of inside attack that can be made by code-protected infrastructure or highly secure systems.

And just as there are increasingly sophisticated ways that distributed denial of services (DDS) can overload websites, they could also certainly be used to disable vital GNSS navigational or timing services. A higher powered simulated but bogus GNSS signal could be used to steer an unsuspecting user into a dangerous or even a fatal activity. Consumers have become so accustomed to relying on the accuracy and reliability of GNSS signals and navigational instructions that they are in no way prepared for a potential criminal or terrorist attack that might be undertaken by either the jamming of a GNSS signal or a counterfeit signal that simulates signals from a GNSS network [11].

The latest GPS Block III satellites, for instance, have been designed to not only be more resistant to jamming and fake bogus signals but to be invulnerable to potential cyber-attack. The challenge is to have tools that aid the automatic detection of signal interference, jamming or spurious GNSS signals.

Vulnerability of Automated Systems That Depend on GNSS Systems and Backup Options

There are several concerns about the ongoing reliability and continuity of service of the various GNSS systems that are now deployed in space. At the most basic level there is the possibility that the battery that is supporting a mobile GNSS device, perhaps in a cell phone, loses power or a chip set might fail.

Then there is the special case of an autonomous vehicle that perhaps comes to a complicated construction site where flag men are diverting traffic into a confusing system defined by cones or a temporary alleyway.

Alternatively there could be a situation with a long tunnel where an autonomous vehicle has lost contact with the GNSS network and needs backup support. In such cases there could be a need for the autonomous vehicle to signal a problem. A possible solution could be provided by a remote 'phantom driver' to respond to such an alarm. The remote human driver equipped with cameras and GNSS systems could then take over control of the automobile, truck or bus to provide temporary navigation through the obstacles until the complication to the autonomous control is resolved. The case of an overlong tunnel excursion or traffic jam in a tunnel could potentially defeat this remote phantom driver's rescue solution unless all tunnels were to be equipped with broadband video mobile telecommunications capabilities. A company known as Phantom Auto is actually testing various telecommunications-based capabilities to allow a backup human driver to assume control of stranded autonomous vehicles that find themselves in a confused or unsafe situation where the onboard computer system shuts down the autonomous controls [12].

The more serious problems would be with the GNSS satellite system itself. There are several catastrophic events that could have a calamitous effect on the half dozen GNSS networks now in operation. One major concern would be a coronal mass ejection (CME) on a particularly violent scale that moves in the direction of Earth, perhaps similar in size to the Carrington event of 1859. Or there could be an atomic blast detonated near Earth that creates a strong electromagnetic pulse (EMP) that would also destroy the electronic control systems on GNSS satellites. In both cases the effects would be similar. Yet another possibility would be a particularly strong solar radiation flare that would be of sufficient magnitude to harm the satellites' electronic control systems.

The problem today is that the GNSS networks now in operation around our planet have become such critical infrastructure that the loss of these systems could have a catastrophic effect on the economy and transportation safety. These systems now provide truly vital services such as synchronization of the Internet and the use of these systems for the takeoff and landing of aircraft. They are also used for time stamping of banking operations and a wide variety of security activities. Perhaps in the next few years these GNSS networks will also become keys to autonomous vehicular self-driving cars and trucks. There are now various short videos available on You Tube with titles such as "Cosmic Hazards" and "If there were a day without satellites" that provide dramatic examples about how vital these networks have become to the entire global economy. They explore the consequences if the GNSS systems were to be disabled by solar storms, a manmade EMP or perhaps via a cyber-attack.

Care must therefore be taken as we move forward to design more redundancy into the GNSS networks and ground units to ensure with greater certainly their longer-term sustainability. This means user units and especially smart phones must be able to access more than one of these satellite networks if for some reason there should be a failure in one of these networks.

The Global Proliferation of GNSS Systems

The advent of more and more GNSS systems has continued to move forward. At one point there was just the GPS and the Glonass systems. These systems were deployed for military reasons to support the precise tracking and guidance of ballistic missiles. The military purpose of these networks as well as the provision in these satellites for "selected availability" to obscure the exact positioning and navigation abilities of the GPS network led other countries to design and deploy alternative GNSS networks around the world.

Thus there are now six GNSS networks operating in orbit or now being implemented. These are the U. S. Navstar Global Positioning Satellite network, the Russian Glonass network, the Chinese Beidou network, the Japanese quazi-zenith network, the Indian Regional Navigation Satellite network, and the European Galileo network, which is currently being deployed. The United Kingdom has indicated that it may feel compelled to design and implement its own network if a cooperative agreement is not reached to allow its full participation in the Galileo GNSS network.

And this is not the only possibility for additional new GNSS networks. The Republic of Korea has indicated that it has plans to design, build and launch a new regional GNSS system. It will be phased in over time and is currently projected to be fully operational as of 2034. The South Korean Ministry of Science and Information and Communications Technology (ICT) has announced that it will take the lead in creating the Korean Positioning System (KPS).

The current lengthy schedule is to develop a ground test of the system components in 2021, design the core elements of the satellite navigation technology in 2022 and begin the production of the KPS satellites as of 2024. The current proposed design of the system (that might ultimately be different) would be to launch and operate a total of seven navigation satellites for the regional system, with three of the satellites being deployed in geostationary orbit above the Korean Peninsula. The effective navigation, timing and positioning system coverage area is to be up to a 1,000 miles around the perimeter of South Korea. The current navigational requirements are to utilize the U. S. GPS system, but the problem is that North Korea has frequently jammed the signals from it [13].

The question thus arises as to how many GNSS networks make technical, operational, and economic sense in our modern world? Does it make sense to have quite so many of these systems, especially since they essentially undertake the same function?

Other satellite applications such as satellite communications, satellite broadcasting, remote sensing, etc., provide unique services with increasing value to their users by relaying individualized messages or capturing images of particular sites. Additional spacecraft to provide more communications or remote sensing thus adds additional value, but satellite navigation systems are redundant and only add value in the event of spacecraft failure of the other systems. One could thus argue that adding more and more GNSS systems only increases the potential hazard of creating new orbital space debris and represents needless investment. In short, adding a fifth, sixth, seventh and eighth new GNSS system does not reasonably represent what can be described as true value once a reasonable level of redundancy and backup is achieved. Super redundant systems do not allow new applications or uses, does not improve accuracy or efficiency of operation and uses up more RF frequencies, orbital space. In time it could also lead to additional orbital debris.

Conclusions

There have been a number of efforts to use space systems to create exact navigational systems from early in the space age. The first such systems used techniques such as Doppler shift that occurred as a satellite flew overhead in a precisely known orbit. The latest system that uses propagation times associated with transmission from satellites with ultra-exact atomic clocks on board is by far the most exact. With the latest atomic clock technology and calculation capabilities with the most exact chip sets in ground GNSS units, the level of precision has dropped from meters accuracy to even centimeters or even millimeters. This level of precision is sufficiently sophisticated to support targeting of missiles, the most exact surveys, or scientific geodetics work.

The current progress in the field of GNSS satellite navigation and precise timing is thus focused on essentially three areas. One area of concern is to work around and prevent the ill effects of jamming of GNSS signals coming from orbiting satellites. The second and closely related concern is to provide for the security of GNSS operations in order to prevent hacking. These cyber-security efforts seek to defeat attempts to hack into navigational or timing systems in either space or on the ground as well as in the military or commercial air space below outer space.

There are new cyber-security systems being developed to limit or prevent any and all adverse effects that might occur by virtue of interference with transmissions from GNSS networks. There is a particular effort to avoid distorting the accuracy of military tracking and targeting systems, and all types of automated systems that rely on GNSS systems such as aircraft avionics or autonomous vehicle navigation and control.

The third area for development with regard to GNSS operations is on the other side of the coin. These are research and development efforts in new applications, services and uses that depend on the effective use of GNSS systems. These efforts also include programs to coordinate cooperative use of all of the space-based systems. This activity would include efforts to create global standards, allocation of frequencies and new regulations with regard to anti-jamming measures, or other measures that would benefit all potential users all over the world. These new applications for GNSS networks can be of enormous value to new areas such as space traffic management, space situational awareness, etc. It might prove useful to the future safe utilization of the protozone, which is above commercial airspace. These new applications might include such activities as safe hypersonic transportation by means of space planes, the secure operation of high altitude platform systems, navigational support for robotic freighters operating above commercial air space and even the operation of dark sky stations.

References

1. Pelton, J.N., Camacho-Lara, S.: Introduction to satellite navigation systems. In: Pelton, J.N., Madry, S., Lara, S.C. (eds.) Handbook of Satellite Applications, 2nd edn, pp. 723–734. Springer Press, Cham (2017)
2. Camacho-Lara, S.: Current and future GNSS and their augmentation systems. In: Pelton, J.N., Madry, S., Lara, S.C. (eds.) Handbook of Satellite Applications, 2nd edn, pp. 781–790. Springer Press, Cham (2017)
3. Continuously Operating Reference Station. https://geodesy.noaa.gov/CORS/. Last Accessed 19 June 2018
4. Op cit, Sergio Camacho-Lara, pp. 781–804
5. Lin, J.: China's future satellite navigation will be millimeter-accurate. Popular Science, Nov. 7, 2017. https://www.popsci.com/china-beidou-3-satellite-navigation-system
6. Enriquez-Caldera, R.: Global navigation satellite systems: orbital parameters, time and space reference systems and signal structure. In: Pelton, J.N., Madry, S., Camacho-Lara, S. (eds.) Handbook of Satellite Applications, 2nd edn, pp. 735–762. Springer Press, Cham (2017)
7. GNSS: Market Report (2015) http://www.navipedia.net/index.php/GNSS_Market_Report
8. Satellite Industry Association-Tauri Group: State of the Industry Report. Washington, D.C. (2016)
9. Hvistendahl, M.: You are a number. Wired Magazine, January 2018. pp. 48–55
10. Myers, B: Secret scores. Wired Magazine, January 2018. p. 56
11. Pelton, J., Singh, I.: Digital Defense: A Cyber Security Primer. Springer Press, Cham (2015)
12. Burke, K: A designated driver for the driverless car: Phantom Auto tests remote-control system Feb. 17, 2018. http://www.autonews.com/article/20180217/MOBILITY/180219775/phantom-auto-tests-remote%20control%20for%20driverless%20cars
13. Korea will launch its own satellite positioning system, GPS World, February 5, 2018. http://gpsworld.com/korea-will-launch-its-own-satellite-positioning-system/

5 The New Capabilities of Weather Satellites

Introduction

Weather, or meteorological, satellites are in fact essentially remote-sensing satellites. What is an important distinguishing feature about this type of satellite application is that they are essentially all operated as governmental services that are supported by tax revenues rather than commercial revenues and services.

The economic challenge is that everyone wishes to benefit from the vital services that meteorological satellites provide, but there is no particularly viable way to offer this service on a strictly commercial basis. This type of satellite is representative of the classic problem in economics referred to sometimes as the lighthouse dilemma. It has always been the case that all ships at sea would like to have lighthouses to warn them of dangerous reefs, yet individual ship owners do not have the wherewithal to pay to build the lighthouses themselves. Thus many functions, whether it is lighthouses, construction of roadways and dams or supply of water have ended up as being offered as a governmental service.

In the United States, however, where private space services are under new analysis, there is consideration as to whether new commercial initiatives might possibly be used in the form of hosted payloads, or might be offered on a commercial basis in other ways. Thus while these type services might not be fully commercialized, as is the case with communications satellites, there may be future prospects to purchase some forms of meteorological data, provided on a supplementary basis by commercial systems.

Everyone one wishes to know about the weather. They especially want to be warned about violent and perhaps deadly storms and even about deadly solar storms such as dangerous coronal mass ejections that can damage power grids, pipelines and satellites. But again individuals are not in a position to pay for expensive weather satellites that today serve several key functions. Those functions are to monitor weather and give warning about dangerous storms, to monitor solar weather and dangerous solar storms, and to monitor climate change and provide warning against

J. N. Pelton, *Space 2.0*, Astronautical Engineering,
https://doi.org/10.1007/978-3-030-15281-9_5

various dangers associated with the longer-term effects of a changing climate.

The development of meteorological satellite technology has been on a continuous upward curve of technological innovation for a half century. The satellite sensors have greatly improved in their ability to monitor weather patterns, to track lightning strikes and storm movement via global lightning mappers (GLMs), X-ray detectors and sounders. These devices are able to operate with great precision to aid in storm monitoring and to detect the direction of storms in virtual real time. And over time additional capabilities have been added. Beyond monitoring and prediction of weather and weather-related events, these now include: monitoring of climate change, monitoring of space weather and especially solar radiation and coronal mass ejections, monitoring of Earth's magnetosphere and even emerging communications for downed pilots, stranded ships at sea and support for search and rescue operations.

Further, there are now both low Earth orbit satellites and geosynchronous satellites that are deployed to work in concert to provide improved data via integrated global imaging. These integrated systems allows better short-, medium- and longer-term weather forecasting. In addition to civilian meteorological satellite services, which are globally shared and coordinated by nations that operate these systems, there are several satellite networks that are operated to support defense and military operations as well.

As more and more countries have developed and orbited their own meteorological satellite systems, this has led to increased coordination and sharing of information. Global cooperation is reflected in many ways, including the sharing of meteorological observations and technical data by means of the World Weather Watch (WWW) that was established through the good offices of the World Meteorological Organization (WMO). This cooperation has also been aided by the Committee on Earth Observation Satellites. Finally, the various space agencies that support the development and in some cases the operation of meteorological satellites have cooperative contingency agreements to insure the continuous flow of satellite images to allow global weather forecasts. Perhaps most notable in this respect was the agreement between Japan and the United States for the GOES-9 satellite to be leased to Japan on an interim basis [1].

The many countries that now operate meteorological satellite systems include the United States, Japan, China, India, Russia, South Korea and those of the European Union. As the effects of climate change have become better understood, additional capabilities have been added to meteorological satellite designs in order to be able to track information related to longer-term trends and to correlate meteorological data. This applies especially to "essential climate variables" (ESVs) and changes to El Nina and El Nino. There is a great deal of additional information collected via meteorological or remote-sensing satellites that we do not necessarily think about yet it is also of critical importance. This is information related to ocean levels, ocean temperatures, land and atmospheric pollution levels, ice and glacial coverage and vegetation cycles over time.

Important New Meteorological Tracking Capabilities

In many ways the optical instruments, CCDs, infrared sensors and radiometers used in remote-sensing activities are largely also common to those used on meteorological satellites. There are some new sensing devices that bring special new capabilities to the latest meteorological satellite networks. For example, the U. S. NOAA satellites have added some special new capabilities in 2017 and 2018.

One of the new capabilities that have been added to the most recent satellites such as the NOAA GOES R, T and U satellites are lightning mappers. These systems are able to determine by means of two new types of sensors the frequency and patterns of lightning strikes to determine the direction, speed of movement and intensity of storms with a much greater precision. They can also be used to aid transportation systems – particularly the safety of aircraft flight and pilot decision-making during storm conditions.

New instrumentation for lightning mapping that is being deployed within the GOES-R, T and U spacecraft includes both an Advance Baseline Imager and the Geostationary Lightning Mapper. These will provide significant economic savings and new levels of safety in terms of storm warnings and prevention of accidents in air and other transportation systems (see Fig. 5.1).

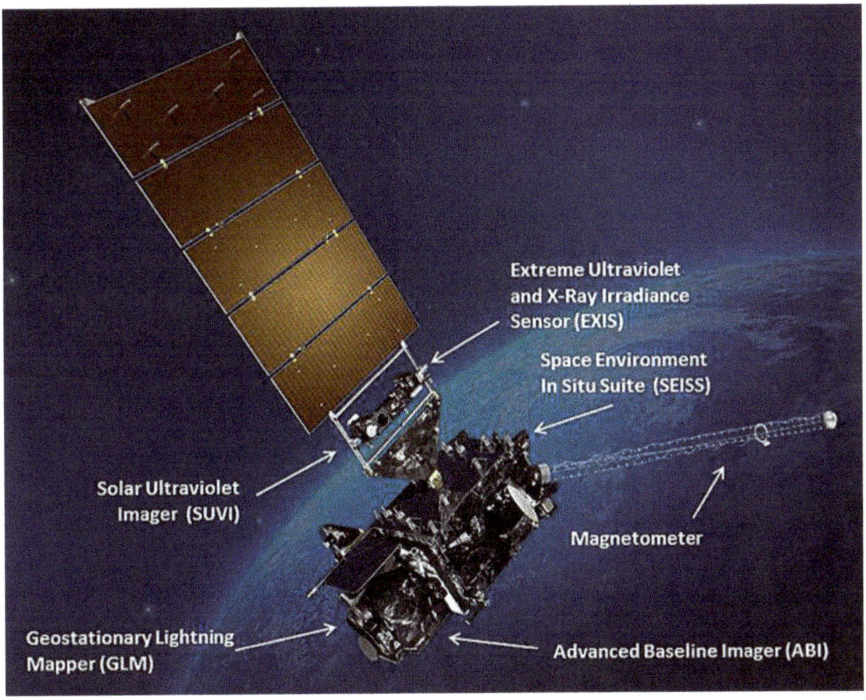

Fig. 5.1 The NOAA GOES-R satellite that features the new GLM and ABI sensors. (Illustration courtesy of NASA.)

The Advanced Baseline Imager (ABI)

The ABI is the primary instrument on GOES-R, T and U for imaging Earth's weather, climate and environment. The ABI on these satellites will provide coverage over a broad range of 16 different spectral bands. These bands include two visible channels, four near-infrared channels and ten thermal infrared channels. The ABI provides more spectral band coverage and higher spectral coverage than earlier types of imagers used on other meteorological satellites. The ABI has two main scan modes that can vary from full disk coverage to particular areas on demand with 30-second updates. Two new capabilities that come with the ABI improved sensing will be the ability to predict the likelihood of fog and its probable density as well as updates on changing storm conditions.

The ability to provided updates for particular targeted areas should greatly increase rapid-time response for coverage at special weather events that can range from thunderstorms, hurricanes, tornadoes, cyclones, to waterspouts. It can also provide targeted updates that relate to fires, volcanoes, or even areas affected by earthquakes. The other new feature that is included in the design is what is called the Geostationary Lightning Mapper (GLM) [2].

Geostationary Lightning Mapper (GLM)

The technical characterization of the Geostationary Lightning Mapper (GLM) is a near-infrared optical transient detector and imager. The GLM is capable of mapping lightning both for in-cloud and cloud-to-ground activity. The GLM is capable of providing a rapid indication of the strengthening of storms and initiating alerts concerning severe weather events. This capability is particularly able to provide improved tornado warnings with at least 20 to 30 minutes advanced notice. In addition, the lightning frequency data that is collected over a period of years can be processed to develop longer-term trend analysis that can be used in studies of climate variability.

It is anticipated that GLM data will have immediate applications to aviation weather services, climatological studies and severe thunderstorm forecasts and warnings. The GLM will provide information that is useful to climate change studies to help note changes in the number and magnitude of both destructive thunderstorms that occur both over land as well as in ocean areas. The U. S. National Oceanic and Atmospheric Administration (NOAA) has attempted to project the potential benefits of these satellites over their lifetime in terms of the warning and preventive precautions value of the ABI and GLM instruments and have indicated that these could be as much as $5 billion per satellite over the length of their in-orbit lifetime [3].

GLM measurements can provide vital information to help with operational preparedness and danger alerts for severe weather. These satellites support aviation, shipping, lightning strikes, fire monitoring and natural disaster reconnaissance.

These new capabilities will be seen in the meteorological satellite systems of many countries in future years. These advantages have been catalogued in the author's *Handbook of Satellite Applications* (2nd Edition), including at least the following features:

- An increased ability to develop short-range forecasts of heavy rainfall and flash flooding.
- New capability to provide near-real time detection of enhanced lightning activity that with associated improved models, can predict changes in the intensity change of tropical storms, hurricanes, and cyclones.
- Related improved warning capabilities for tornado and severe thunderstorm in terms of increased lead times as well as a corresponding reduction in false alarms and spurious information. (This particularly valuable new capability is of importance for storm warning and transportation routing for oceanic regions, mountain areas and areas where there might be radar outages.)
- Improved routing of commercial, military and private aircraft over oceanic regions, mountain areas and sparsely populated and remote areas during severe storm conditions.
- More accurate and timely warning of lightning ground strike hazards.
- Development of improved and more accurate numerical weather prediction models and increased identification of deep atmospheric convection patterns.
- Increased capability to develop what might be called "lightning climatology" and models of lightning intensity within storms.
- Improved ability to monitor and create mathematical models of a wide range of storm and lightning intensity patterns.

These capabilities, when combined with the capabilities on polar-orbiting satellites in low orbit, such as the U. S. POES and JPSS, provide an expanded range of abilities to: (i) forecast and better model weather events and climate change; (ii) assist with location of and recovery from natural disaster-related events, and (iii) help industry and communities to minimize the adverse effects of weather and natural disaster events by avoiding their worst effects and through better forecasting and prediction.

Monitoring of Terrestrial Gamma Ray Flashes (TGFs)

One of the unexpected benefits of lightning monitors on-board meteorological satellites has been a better understanding of terrestrial gamma ray flashes. It was at one time assumed that gamma ray flashes came either from the Sun or from cosmic sources at astronomical distances. The X-ray detectors, sounders, radiometers and spectrometers on-board meteorological satellites have confirmed that lightning flashes can give rise to X-rays and gamma radiation bursts. (See Fig. 5.2.)

The satellites that detected these types of very high energy events were never particularly designed for this purpose, and the more recent discoveries concerning this phenomenon have come from other sources. One of the most accurate sources of information about the characteristics of the very short-lived phenomena is the so-called Telescope Array, which is an observatory that is composed of 507 scintillation surface detectors. These detectors, which act as a sort of radio telescope,

Fig. 5.2 Lightning can release microsecond-long terrestrial gamma-ray flashes. (Graphic courtesy of NOAA.)

are spread out across a large distributed expanse of the Utah desert area. A report on the study of terrestrial gamma-ray flashes reports the following: "TGFs were originally measured by satellites, but the observations were imprecise. Thanks to the Telescope Array data, scientists now know the bursts last for a few dozen microseconds or less. The new data also proves TGFs originate in the thundercloud near the initial breakdown stage of the lightning" [4].

New Satellite-Based Monitoring of Essential Climate Variables (ECVs)

The earliest meteorological satellites were focused on weather imaging and forecasting, but it was soon discovered that the cumulative data from these satellites could aid understanding of climate trends. As satellite sensors grew in sophistication and as scientific precision grew, more and more capabilities were added to meteorological satellites. One of these important expanded capabilities was the addition to meteorological satellites of a sensor that could help to accumulate data related to monitoring climate change and to interpret longer-term shifts in the world's climate.

The deployment of satellites that were primarily designed to help predict weather did not always produce the data that was considered most vital to understanding longer-term effects related to climate change and obtaining key climatological data. Thus there have been a number of research satellites developed and launched by space agencies around the world. Appendix A in this book provides information with regard to some 34 satellites that have been launched to aid research into longer-term climatological changes. In some cases these satellite have played a dual role of providing operational data as well as research information and insight [5].

Keys to climatological research are two main factors. One factor is having data over a significant time period sufficient to identify longer-term trends. There are now in a number of cases data from observation satellites for a period of some 50 years. The second factor is to be able to assess the accuracy, reliability and quality of data from satellites, especially if taken from different satellites with different sensors that have variations in their design, sensitivity or calibration.

For the purpose of providing accurate and comprehensive climate monitoring data, a program known as SCOPE-CM was established a decade ago in 2008. SCOPE-CM stands for Sustained and COordinated Processing of Environmental satellite data for Climate Monitoring. SCOPE-CM represents a network of agencies and operators of environmental satellite systems that includes EUMETSAT, the Chinese Meteorological Administration, the Japanese Meteorological Agency and the U. S. National Oceanic and Atmospheric Administration, which also provides research data from NASA.

A number of key international agencies participate in this program. They include the World Meteorological Organization (WMO), the specialized agency of the United Nations, the World Climate Research Programme, which is a part of the World Climate Programme of the WMO, the Global Climate Observation System of the WMO, the Coordination Group for Meteorological Satellites (CGMS), the Committee on Earth Observation Satellites (CEOS), the Committee on Space Research (COSPAR) and other international groups that play a key role in climate observation. The purpose of SCOPE-CM is to support, coordinate and facilitate international activities to generate comprehensive and verifiable climate monitoring data from multi-agency satellite data.

Within SCOPE-CM, there are processes to develop data with respect to specifically identified climate monitoring categories that include such areas as carbon dioxide levels, methane levels, other greenhouse gas levels, sea level rise, desertification, global temperature increase, lakes, icecap coverage, etc. SCOPE-CM facilitates cooperation on the basis of shared and distributed responsibilities for the generation of global climate monitoring and essential climate variable products [6].

Closely associated with this process is the global initiative that is known as WG Climate. This activity has now created an effective mechanism for global sharing of climate-related data. The full name of WG Climate is "CEOS/CGMS Working Group on Climate." [7]

Solar Activity Monitoring and Dashboard Display in Near Real Time

The initial systems such as the NASA developed TIROS and Landsat satellites were focused on finding ways to use spacecraft to monitor weather conditions and carry out Earth observation. NASA, however, in its research missions also sought to carry out reconnaissance and research programs for the planets and the Sun. The costs of the Space Transportation System (STS) and the International Space Station (ISS) led to cutbacks in planetary and solar research projects. This not only resulted in the development of improved meteorological satellites but also promoted the idea of putting radiometers and other sensors on these spacecraft to monitor the Sun and solar weather as well as Earth weather.

Thus the meteorological satellites that NASA developed for NOAA increasingly included solar wind and radiation sensing capabilities. This pattern was also seen in the meteorological satellites that the space agencies developed in Russia (Meteor), Japan (GMS, Himawari) China (FengYun satellites) and India (Insat) with Sun-sensing capabilities.

The latest GEOS satellites, as represented by GEOS-R and shown in Fig. 5.1, contain a Space Environmental In Situ Suite (SEISS), a Solar Ultra-Violet (SUV) Imager and an Extreme Ultra-Violet and X-ray Irradiance Sensor (EXIS). These sensors allow NOAA to update a dashboard display on solar wind and alerts with regard to coronal mass ejections and solar flares. In fact the U. S. National Weather Service provides dashboard displays that include solar weather forecasts and near real time information that is specific to the areas of aviation, electric power, emergency management, global positioning systems, radio, satellites and even space weather (see Fig. 5.3) [8].

Improved Monitoring of Changes to Earth's Magnetosphere

Yet another aspect of meteorological satellite monitoring capabilities comes with respect to gathering data on Earth's magnetosphere. The ability of Earth to withstand massive solar weather storms actually depends on the world's magnetosphere. Earth's magnetic poles form a massive protective magnetic shield around the world. This is what shapes the Van Allen Belts. The magnetic shielding diverts ions from the Sun that travel at over a million kilometers per hour when spewed from the Sun during a coronal mass ejection. If it were not for this natural magnetic shielding the incoming high-energy ions would blast Earth's atmosphere and knock out electrical power systems and other infrastructure that range from pipelines, communications systems, to industrial control systems, to other satellites. When coronal mass ejections are forecast to hit Earth, satellites, electrical power systems, pipelines and many networks that depend on electrical energy power down to avoid massive outages. This problem and protective strategies against solar flares and coronal mass ejections are addressed in a later chapter.

A number of additional capabilities to study Earth's magnetosphere, such as the NASA's MagnetoMMS satellite mission, with a constellation of four satellites [9], and the ESA Swarm with a three-satellite constellation [10], are currently seeking to map the current shift in Earth's magnetosphere. These satellites have confirmed that a key shift in the magnetosphere is currently in progress. Magnetic north has now moved down to Siberia and magnetic south has moved up toward New Zealand.

This magnetospheric mapping exercise suggests that Earth's electronic grids around the world could be increasingly vulnerable to a major coronal mass ejection in coming years, as the natural

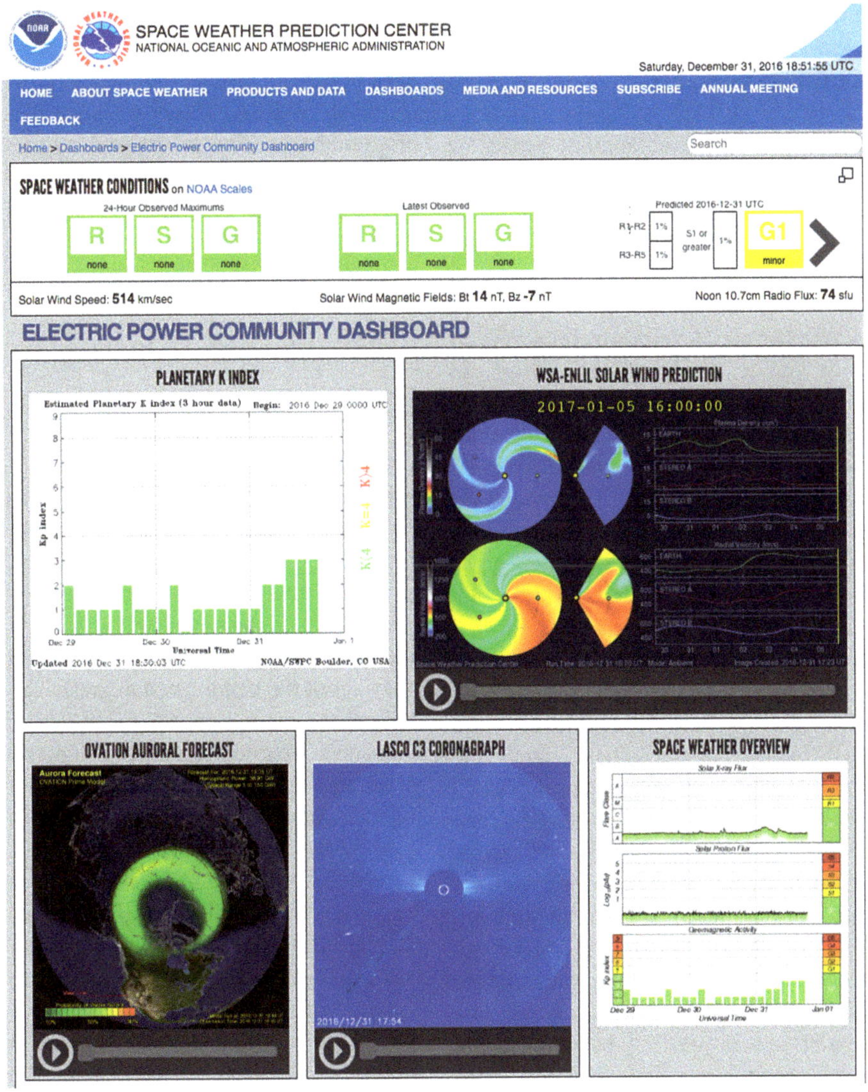

Fig. 5.3 NOAA National Weather Service dashboard display for electrical power. (Graphic courtesy of NOAA.)

protective shielding is reduced to a significant degree. If there should be a massive strike, similar to the Carrington Event of 1859, or even more recent events such as the Montreal Event of 1989 or the Halloween Event in Scandinavia of 2003 while Earth's natural shielding is down, the economic and infrastructure destruction consequences could be of catastrophic proportions and might require many years to recover from.

Emergency and Search and Recovery Communications

Finally, there is yet another aspect of meteorological satellite functionality, and this is to aid with search and rescue operations. GOES series satellites carry a payload supported by NASA's Search and Rescue (SAR) office, which researches and develops technologies to help first responders locate people in distress worldwide, whether from a plane crash, a boating accident or other emergencies. NASA has developed emergency beacons that are placed on most aircraft, ships and many vehicles operating in remote and perilous conditions. These beacons are automatically activated when an aircraft goes into the water or on severe impact.

There are three types of distress radio beacons that can be registered with COSPAS-SARSAT. These are the Emergency Location Transmitter (ELT) beacons for aviation use, the Emergency Position-Indicating Radio Beacons (EPIRBs) that are used for maritime applications, and the position location beacons (PLBs) for personal use. One of the great mysteries about the missing Flight 370 Malaysia aircraft is why the search and rescue ELT beacon did not work when it crashed. It should have immediately been activated and started transmitting as soon as the aircraft hit the water, and continued transmitting for a number of days.

The GOES is a geosynchronous satellite, and the GOES series thus has a global vantage point of the entire world. There are beacon relay systems on low Earth orbit satellites as well that can help target the site of a disaster with greater precision.

Many hundreds of crashed pilots and stranded crews and passengers on boats have now been rescued by means of the COSPAS-SARSAT beacons since this program began. This program also relies on the GNSS position location capabilities of the various satellite networks that participate in this program, which includes the U. S. GPS system and the Russian C [11].

Conclusions

The importance of meteorological satellites to the world economy as well as to human safety cannot be underestimated. The fact that these satellites provide a vital service in so many different areas is, however, not so widely understood. There is the apocryphal story about the Congressman that asked why there was a need to purchase expensive weather satellites when one could just turn on the weather channel to find out the latest weather conditions. Most people understand that weather information today comes from a combination of GEO and LEO satellites, aircraft and even drones and land-based instruments, but they likely are unaware of the sophisticated combination of sensors that provide images in the light spectrum as well as in the infrared up to the X-ray spectrum to understand weather conditions and the most severe of storms. Meteorological satellites can monitor lightning strikes, solar storms and flares, gamma ray events, help measure Earth's magnetosphere and relay signals from emergency search and rescue beacons.

Today a wide range of satellite activities and services have been commercialized, and their services in communications, networking, remote sensing, etc., represent a part of the global economy. Up to this time meteorological services have been provided as a governmental responsibility, but in the United States, at least, there is the possibility that some of these services might in the future be offered as hosted payloads on commercial satellite missions.

See the Addendum to this chapter in Appendix B at the end of this book for a list of current, past and proposed research satellites for climate-change purposes.

References

1. Hinsman, D.: Implementation Activities within the WMO Space Programme. World Meteorological Program (2016)
2. GOES-R Spacecraft. https://www.goes-r.gov/. Last Accessed 23 June 2018
3. Camacho-Lara, S.: United States Meteorological Satellite Program. In: Pelton, J.N., Madry, S., Camacho-Lara, S. (eds.) Handbook of Satellite Applications, 2nd edn. Springer Press, Basel, p. 1185 (2017)
4. Hays, B.: Study confirms link between gamma rays, lightning strikes. UPI. Com Science News. https://www.upi.com/Science_News/2018/05/17/Study-confirms-link-between-gamma-rays-lightning-strikes/8261526583038/ (17 May 2018)
5. List of Cimate Research Satellites. https://en.wikipedia.org/wiki/List_of_climate_research_satellites. Last Accessed 25 June 2018
6. EUMETSAT generates Climate Data Records for NWP model-based reanalysis, climate system analysis and climate modelling applications. https://www.eumetsat.int/website/home/Data/ClimateService/index.html. Last Accessed 25 June 2018
7. WGClimate The CEOS/CGMS Working Group on Climate. http://ceos.org/ourwork/workinggroups/climate/. Last accessed 26 June 2018
8. NOAA Space Weather Conditions: Current Space Weather Conditions. https://www.swpc.noaa.gov/dashboards. Last Accessed 26 June 2018
9. NASA Magnetospheric Multiscale (MMS) Satellite. https://mms.gsfc.nasa.gov/about_mms.html. Last Accessed 25 June 2018
10. Earth OnLine: What is Swarm? https://earth.esa.int/web/guest/missions/esa-operational-eo-missions/swarm. Last Accessed 25 June 2018
11. Satellites - Search and Rescue Satellites. www.satellites.spacesim.org/english/function/search

6 New Uses of the Protozone

Introduction

There is an interesting book by David Loth entitled *How High Is Up* that explored the problem that has been an issue for national armed forces seeking to defend their countries for many centuries. The conclusion in Loth's book is that despite efforts to define things like national air space and sovereignty over land of a particular country, the practical answer has been the area that can be effectively defended. Today, national commercial air space is generally accepted to rise up to 20 kilometers (12.5 miles).

Military air space extends well above commercial air space and is assumed by many to rise all the way up to where outer space begins. Some have suggested that outer space starts at the Von Karman line, where planes can no longer fly, or 100 kilometers (62.5 miles). Some, such as Russia, have suggested that this is 110 kilometers, while others have suggested that it is 160 kilometers, which is the altitude where a satellite can remain in its orbit on a truly sustained basis. Yet other countries extend their claim out to the geosynchronous orbit and even beyond.

One of the reasons that defining what is national air space and what is outer space has not proved to be an issue of major practical importance is that there have not been many significant applications for the use of the area above commercial air space and below outer space where satellites can sustain their orbits. Jonathan McDowell at Harvard University has done an interesting study of various satellites that have maintained orbit for brief periods of time in the altitude ranges of about 100 to 160 kilometers, but this is largely an academic study, rather than being definitive in establishing exactly where outer space begins [1].

However, the rise of Space 2.0 and new thoughts about using the subspace, or the protozone (i.e., the zone above 20 and below 160 kilometers) for a number of practical purposes has increasingly given rise to concerns about who might use this near space area and under what regulatory or licensed control authority? In short the prime question has been as

J. N. Pelton, *Space 2.0*, Astronautical Engineering,
https://doi.org/10.1007/978-3-030-15281-9_6

to what purposes might it be safely used and under whose authority or control? Is there ultimately going to be a mechanism for the legal, administrative, safety, and practical control of this region? If there is ever a systematic global approach to space traffic management, does it also apply to the protozone area? These questions are just some that Space 2.0 gives rise to and which no one has ready answers.

There are now perhaps a dozen possible uses of the protozone that have arisen. These uses include high altitude balloon and dirigible ascents for tourism and other purposes; telecommunications or IT services; surveillance or weather monitoring; high-altitude platforms that might also be used for telecommunications, IT, Earth observation or other purposes; high-altitude robotic air freighters to carry shipments at lower cost and higher efficiency; dark sky stations for research and to allow small payloads to be flown to low Earth orbit using electronic propulsion; high-altitude hypersonic flights for space tourism or transport: and various types of hypersonic vehicles or platforms that could be deployed for offensive or defensive military purposes. This might include such weapons as scram-jet missiles operating at speeds of Mach 6 to Mach 8.

In the last few years there has been increasing concern and interest in the issue of space traffic management and space situational awareness, in order to make operations in Earth orbit safer and more systematically controlled. This discussion has generally not addressed at what altitude space traffic management would begin and whether the protozone would be a part of this new international management and control system.

Space Situational Awareness

There are essentially two ways that might be used to monitor and thus keep track of orbit space debris, operational spacecraft in Earth orbit, and objects in the protozone. One means is via the U. S. S-band radar tracking system recently deployed in the Pacific Ocean region, and the other way would be to have GNSS units onboard spacecraft that are optimized for Earth orbit operations and are equipped to report on their exact location in near real time. This would, of course, work only for currently operational spacecraft and provide information for space debris that can only be tracked in a passive manner.

The new billion-dollar S-band Space Fence, that will be able to track some 22,000 space objects in low and median Earth orbit, is a facility that has been installed in the Pacific Ocean area on Kwajalein atoll by Lockheed Martin under a contract with the U. S. Air Force between 2015 and 2018. This facility has the sensitivity to track objects down to 10 centimeters in diameter in low Earth orbit (LEO). (See Fig. 6.1.)

A further S-band radar tracking facility has been discussed that would be located in Australia. This new facility would replace the earlier VHF-band radar system in Texas and Arizona known as "the Fence" and has a tracking sensitivity that is ten times greater than the earlier facility does [2].

Other key new radar systems include the DLR-funded GESTRA facility, which uses a new phased-array antenna design and is also known as the TIRA system. This facility will be able to track debris elements as small as 1 centimeter in size in low Earth orbit and would be

Fig. 6.1 The U. S. S-band Space Fence facility located on the Kwajalein atoll. (Illustration courtesy of Lockheed Martin.)

of great utility in tracking objects in the protozone [3].

Of the 22,000 objects currently being tracked in low or medium Earth orbit, some 20,000 are defunct space objects without active control systems. This represents a large problem in that space traffic management implies the ability to control space objects, but these space debris objects – in the form of spent rocket motors, defunct satellites, etc. – are without means to allow anyone to steer them or actively de-orbit them. Fortunately objects in subspace are often subject to some level of control and in some cases are capable of being steered or maneuvered except for the notable case of balloons. In short, effective space traffic management for the protozone is more likely to be implemented, especially at the national air traffic control level. Prime reasons for this are:

- Affected craft or objects are in lower altitude and thus easier to track and access.
- Many more objects are capable of being steered or maneuvered.
- Responsibility for the control of these objects is more likely to be carried out by experienced national aviation safety air traffic control – at least in most instances.

There are, of course, important exceptions to national air traffic control agencies being able to exercise exclusive traffic control for vehicles or objects in the protozone. Notable issues would include hypersonic vehicles that would be engaged in international transportation or the flight of scram-jet type hypersonic weapon systems.

There have been studies and particular proposals as to how better coordination of space situational awareness and space traffic managements might be carried out, and several of these ideas and suggestions would also likely apply to the protozone [4]. To date these have not been actively implemented, and there are currently no widely agreed processes whereby these could be formally implemented. Further progress in this area is complicated by recent U. S. space policy changes coming from the U. S. executive branch with regard to the possible creation of a U. S. Space Force [5]. Likewise there are potential complications that might come from the signing by the U. S. president of what is known as Space Directive 3. This document, signed by the president, charges the Department of Commerce, in coordination with the Department of Defense, to develop new capabilities with regard to space situational awareness and space traffic management capabilities – among other tasks [6].

New Technologies Under Development

Currently the world of NewSpace is creating not only a wide range of new technologies in terms of new systems and concepts for innovative spacecraft and launch systems, but it is also fueling many new and innovative ideas about how the region just below space might be used. The thought is that systems that do not have to go all the way into space might be more cost effective but still might offer new avenues to provide communications and IT relays, improved surveillance and observation, as well as new possibilities for near-space tourism, higher speed transportation systems or even new types of weapons or defense systems.

This has led to research in such diverse areas as: (i) dirigibles or specialized balloon systems for space adventures; (ii) high altitude transponder systems for communications or Internet connections, especially for improved coverage and interconnection for the many rural and remote regions of Earth that are currently not connected to the Internet; (iii) new types of systems for Earth observation, surveillance, fire detection, criminal activity monitoring, and other forms of remote detection or observation systems; (iv) new forms of hypersonic transportation plus weapons systems, coupled with research into sonic boom suppression; (v) robotic transport; and (vi) extremely high-altitude dark sky stations.

Reports on the status of these various research and development efforts goes well beyond the intended scope of this book, but the possible new applications in these various areas are outlined below. Further, since progress in these areas is currently so very rapid, any attempt to report on these technical areas and the current status of research and development would likely become outdated in a very short period of time. As an alternative, it is suggested that anyone interested in the current state of development in these various areas from a technical point of view might undertake a web search.

The Growing Number of New Applications for the Protozone

It is remarkable how the development of new technology and the passage of time can greatly redefine the value of things. The U. S. purchase of Alaska, and even the Louisiana Purchase, were originally considered highly questionable enterprises. For a great stretch of time near-space, the area below outer space, did not to seem to have any particular purpose or hold any particular value. Today however there are a burgeoning number of new applications, and many people are seeking frequency allocations and other authorization to use this once rather forlorn and almost forgotten area. The following actual or potential uses of the protozone are now in play.

High-Altitude Protozone Tourism

There are currently at least two new companies seeking to develop commercial offerings to clients who wish to experience the thrill and excitement of observing space from a platform that is sufficiently high to see the curvature of Earth and the dark sky, but without the extremely high cost of a suborbital flight and its associated risks. One venture that is the brainchild of former Biospherians Taber McCallum and Jane Poynter is called World View. This company has developed a parafoil that has been able to make ascensions up to 120,000 feet. This is equivalent to 32 kilometers (20 miles). Their business plan is to offer ascensions up to 30 kilometers in altitude with passengers contained in a capsule that is supported by a parachute in case there is a problem with deflation of the balloon system [7].

The other firm, which is based in Spain, is known as Zero 2 Infinity, and it has also had successful test flights of up to 25 kilometers. Eventually it plans to offer tourist flights that will go as high as 36 kilometers (22 miles). These flights will not require space suits but will require special equipment as well as clothing that will not give rise to electrostatic sparks [8] (Fig. 6.2).

Fig. 6.2 World View parafoil that is intended for tourist ascensions up to 30 kilometers in altitude. (Graphic courtesy of World View.)

UAVs or Balloons for Communications, IT and Earth Observation Services and High-Altitude Platforms (HAPS)

The idea is that balloons, unmanned aerial vehicles (UAVs) and high-altitude platform systems (HAPS) might be able to provide wide area coverage for communications, IT, surveillance, or other practical services. Such possible applications have gained great interest in the last two decades.

The small sat revolution has given rise to other ways to find new economies that depart from the old ways that involved expensive satellites and expensive launch vehicle services. Actually a number of ideas have surfaced in this regard. The Japanese first championed new frequency allocations to support communications from high-altitude platform systems (HAPS) and proposed a reverse allocation for the satellite Ku-band allocation. The proposal that was agreed was to have a Ku-band approved use that would be 12 GHz for the uplink and 14 GHz for the downlink, rather than the usual satellite allocation of a 14 GHz uplink and a 12 GHz downlink.

It was Google that came up with the idea that one might send up radio-accessible packages that could fly on balloons similar to meteorological balloons. These might be used to provide connectivity to the Internet in rural and remote areas of the world, where such access was either very limited or non-existent. These so-called "Loons" would carry equipment much akin to the equipment found in a cell tower. One of these tennis court sized Loons, with its radio system suspended below, is designed to allow wireless connection to the Internet in remote areas. Such balloon systems are designed to withstand the wind, cold and ultraviolet radiation conditions of the stratosphere at an altitude of 20 kilometers. They are intended to remain in service for up to 100 days and are then allowed to descend on a controlled basis. There are attempts to create a controlled drift of these balloons so that one balloon drifts off and another replaces it, but this is far from an exact science. They are sufficiently steerable or maneuverable to avoid collisions in the high-altitude stratospheric conditions. Currently the launch and ascension of balloons are controlled by national air traffic control agencies, which started with a release of balloons from New Zealand in 2013. These systems have been used to aid in emergencies such as during recent disasters in Puerto Rico and in Peru [9].

There is even more interest in the use of high-altitude platform systems/stations (HAPS) for telecommunications, broadcasting, surveillance and/or monitoring. Frequencies have been allocated by the International Telecommunication Union (ITU), and additional frequencies are being proposed for the upcoming 2019 ITU World Radio Conference. The current allocations for HAPS service are in the Ku-band (12 GHz uplink and 14 GHz downlink) and possible new allocations in 38 to 39.5 GHz bands globally; and in Region 2 possible additional allocations are in the 21.4 to 22 GHz and 24.25 to 27.5 GHz bands [10].

There have been a number of proposals as to how this type of service might be offered via solar cells and electric-powered dirigibles such as a Japanese system with 17 HAPS vehicles, balloon systems, or long endurance powered aircraft known as High-Altitude Long Operation/Endurance (HALO/HALE) platforms. These HALO or HALE platforms were most commonly seen as being either solar-powered with electric propeller systems or jet aircraft that would constantly cycle in and out of their fixed location. There have even been proposed futuristic concepts that would use beamed microwave power to stabilize these platforms in a steady or relative stable flight path. In short there have been over a dozen such systems proposed over the past decade. These various proposals were largely seen as the next generation of communication platform beyond the technology represented by earlier aerostats.

Most recently Facebook has decided to close down its support for the Aguilar High Altitude Platform System, which it had been backing for some time. Instead of the Aguilar project it is now going to seek partnerships to create HAPS or UAV projects [11]. In this evolving story, the latest chapter is that Facebook is building an in-house test satellite called Athena for launch in 2019. This satellite is also intended to provide Internet services to rural and remote areas, and if the test is successful, then presumably a larger satellite constellation would follow in due course [12].

At this time none of these earlier proposed HAPS projects are now operational, even though the Aguilar project might continue under alternative funding. The typical HAPS project would perhaps be deployed at an altitude of about 20 kilometers (see Fig. 6.3.)

One significant option in the HAPS development field is that Thales Alenia has now developed and tested a craft that they characterize as between a drone and a satellite. They have foreseen many different functions that a HAPS could provide, including telecommunications, broadcasting, Earth observation or monitoring services. (See Fig. 6.4.)

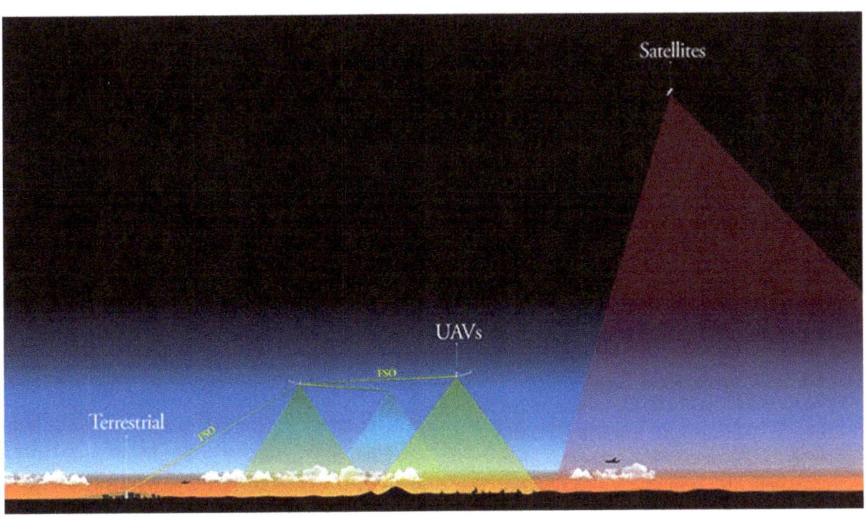

Fig. 6.3 Illustration of relative heights of terrestrial cellular, UAV or HAPS, and low Earth orbit satellites. (Graphic created by the author.)

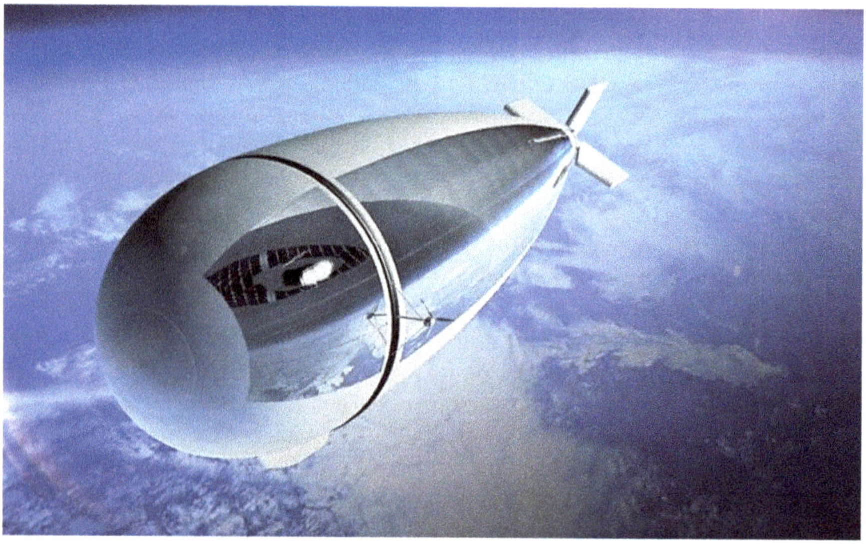

Fig. 6.4 The Stratobus[TM] HAPS as developed by Thales Alenia. (Graphic courtesy of Thales Alenia.)

The area of 17 to 22 kilometers is considered a suitable range of altitudes for HAPS operation, since wind speeds at these altitudes are typically at or below 100 kilometers per hour rates, and temperatures and UV radiation levels are generally acceptable for HAPS stations to be able to perform their functions for long periods of time.

Jean-Pierre Thessel, who heads the development of Stratobus for Thales-Alenia, is enthusiastic as to the possibilities. He has envisioned that this type of platform could be deployed to an altitude of 20 kilometers for many diverse applications that might become possible at a future date. The Stratobus[TM]

> *...could be used for surveillance missions, including land, maritime, oil platforms, piracy at sea, and.... environmental monitoring missions.... [It could] carry both radar and optical imaging payloads, for continuous surveillance capability, day or night and under all weather conditions. For military applications, it can be displaced as theaters of operation move. It would also be very useful in the telecommunications market, to provide 4G and 5G connectivity in the future. Its position in the stratosphere at an attitude of 20 kilometers is an optimal position for 5G, which demands very short latency of just a few milliseconds. Concerning navigation applications, it would strengthen the AIS* [Automatic Identification System] *network in heavy traffic zones, thus improving traffic control* [13].

This type of service via a HAPS station might prove particularly useful to provide cellular connectivity in recovery efforts where local mobile networks have had their power knocked out and for island countries the size of Jamaica, Dominican Republic, Haiti, Aruba, Singapore, Taiwan, etc., where one or a very few platforms could essentially cover the entire nation.

Dark Sky Stations

Another intriguing concept for deployment in the protozone is that of the dark sky station. This would involve the deployment of a lighter-than-air facility in the stratosphere with the purpose of carrying out a range of different missions. These missions might be to conduct scientific, meteorological or other practical experiments and demonstrations within the stratosphere.

These missions for the dark sky stations might include: (i) serving as a locale for short-term occupancy by human engineers and scientists for experimentation involving ozone layer or geomagnetosphere measurements, meteorological data, northern lights or climatological observation, and/or perhaps calibration of experimental gauges; (ii) serving as a launch point for electric ion propulsion-powered small satellites that could, at an altitude of perhaps 20 kilometers, be launched to spiral upward to achieve orbital speed and altitude; (iii) operating as a terminus and final checkout point for the launch of rockets that have been lifted by balloons to this altitude; (iv) providing a location for a new type of space tourism-type experience; and (v) serving as a multi-purpose location for commercial applications such as telecommunications, broadcasting, Earth observation, monitoring, etc.

JP Aerospace of California has undertaken the most concentrated study and experimentation into the possible feasibility. Below is a current prototype version of a dark sky station that JP Aerospace volunteer engineers have developed. But this is only for a proof-of-concept design. (See Fig. 6.5.)

Fig. 6.5 Prototype engineering concepts for a dark sky station. (Illustration courtesy of JP Aerospace.)

JP Aerospace has also developed quite ambitious futuristic concepts that have been developed on a theoretical implementation basis, including such ideas as a "Stratostation." This ambitious concept would have five different components essentially a half-mile in length (800 m) joined in a five-star configuration in order to create a giant dark sky station platform for space tourism and launch operations. It would be 1.6 kilometers in diameter [14].

There are many other possible applications that might come from a dark sky station that have yet to be fully explored. A location above a significant portion of Earth's atmosphere might provide an unclouded view of not only outer space, it also might provide a useful downward view of Earth in terms of observing patterns of pollution and to get a more synoptic view of certain climate change patterns.

Automated Robotic Air Freighters

There is today a very large volume of air freight across the oceans and across continents as well. Some of the largest providers of air freight services such as UPS, FedEx and others have begun to look at ways to provide their services more efficiently and at lower cost. One of the options being considered is the idea of a large air freighter that would essentially be robotically controlled and might fly at higher altitudes and somewhat slower speeds in order to conserve fuel. The idea would be for something like a converted Boeing 747 aircraft to fly at 50,000 feet (80,000 km) on such freight hauls. The aircraft would be mostly automated with only one emergency crew person aboard. This robotic freighter might cruise at, say, 500 kilometers per hour at higher efficiency and with much less wind resistance at the higher altitude. The result might be fuel consumption cut in half, and labor costs might be reduced by a factor of four.

In moving forward with such concepts one must look not only at the advantages, but also potential problems. There are many ideas for higher altitude flights at various speeds for different purposes. It is often not appreciated that the same amount of pollution released at sea level and in the stratosphere does not produce the same result. In a region where the atmosphere might be 10 times, 25 times or 50 times thinner, the impact of pollution is much more severe. In all of the ideas that include high altitude robotic air freight services, hypersonic transportation, suborbital flights for space tourism, hypersonic missile systems and launchers that use solid fuels and release particulates, there must be care given to the problem of air and stratospheric pollution and concern about mitigation processes that might be undertaken to diminish these harmful impacts [15].

There have been rumors for years that the U. N. World Meteorological Organization (WMO) and the U. N. Environmental Program (UNEP) would undertake a major review of air pollution coming from aircraft stratospheric flights, but none has proceeded. Various studies have been undertaken of the polluting effects of solid-fuel rockets in particular as well as other investigations of the effects of upper altitude aircraft and rockets on the stratosphere, but none of these efforts have triggered a global review of what this means to pollution of the upper atmosphere. This must be one of the issues that the U. N. COPOUS Working Group on the Longer Term Sustainability of Outer Space Activities (LTSOSA) should ultimately consider, perhaps in consultation with the WMO and the UNEP.

Hypersonic Transport Systems

If there is truly a massive new commercial opportunity in the new uses of the protozone, it is very likely to lie in the area of hypersonic transport. This would most likely involve the successful development of scramjet technology that uses air-breathing jet engines capable of enormous speeds or rocket engine systems for space planes developed to provide safe suborbital flights from point A to point B on our small planet. There has been speculative talk for decades of such hypersonic spaceplanes that could fly from London, England, to Sydney, Australia, or from New York to Tokyo in a matter of a few hours or so. Often the hype does not include the time for boarding and deplaning from the aircraft and consideration of the time that might be required to cool a spaceplane that has braked from a speed of Mach 6. Nevertheless the prospect of a mode of transport that could move passengers nearly half way around the world in a few hours is exciting if it could be done safely and without causing longer-term environmental harm. Certainly there are those that believe vacuum-sealed tunnels that use ultra-efficient maglev technology or so-called hyperloop systems may be the ultimate answer rather than hypersonic transport via the protozone.

Certainly there are a number of issues to be seriously addressed. These include:

- Development of reliable rocket propulsion or scramjet engines that can be used many times to make such flights reliable, cost-effective and safe.
- Careful study of environmental effects to confirm that it is possible to accomplish a number of such flights on an hourly basis without major polluting effects to the stratosphere.
- Economic and market studies to confirm that such offerings could be viable as a service that can commercially sustain itself.
- Development of new technology to create Quiet SSPs and Quiet HSPs that can avoid the creation of massive sonic booms near airports and cities as aircraft slow to land.

Currently there are a number of parallel technological development programs going on in the world that are essentially at odds with one another.

First, there are a number of quite advanced studies of electronic propulsion for air travel that would provide a service that is considered much more environmentally sound and that are likely to come first from European manufacturers (i.e., Airbus) and European carriers.

Secondly, there are a number of R & D projects underway in the United States, Europe, Japan and China to develop new jet airliners that are able to fly reliably, safely and cost effectively at speeds in the Mach 2 to Mach 4 range that are envisioned to provide air passenger service. Most of these fly with conventional liquid jet fuels or liquid hydrogen and liquid oxygen.

Thirdly, there are also efforts to create spaceplanes. These are designed to fly at speeds mostly in the Mach 6 to

Fig. 6.6 The A2 hypersonic jet proposed by Reaction Engines, Ltd. (Illustration courtesy of Reaction Engines Ltd.)

Mach 8 range, and many of these use hybrid solid-fuel systems that are capable of throttling. These are currently being designed either for suborbital flights associated with space tourism or as perhaps the first stage of a rocket launch system for small satellites. They include the Space Ship Company owned by Virgin Galactic, the now-defunct Swiss Space Ship (S-3) company or the Skylon single-stage-to-orbit vehicle powered by a "Sabre" scramjet engine and developed by Reaction Engines. Reaction Engines has also developed the plans for a spaceplane called the A-2 that would, in theory, provide service between the U. K. and Australia in under 5 hours. (see Fig. 6.6.)

Fourth and finally there are research agencies hard at work to take on the problem of sonic booms. The research is currently focused on telescope needle noses that could extend from the plane in such a way as to create a series of much smaller booms over time rather than one giant boom as the entire plane makes the transition from supersonic speeds to a velocity under Mach 1.

Anyone who feels that they have a complete and accurate idea as to the final outcomes and practice of commercial services actually to be provided say five to ten years from now would likely be mistaken. In fact there are many different and quite complex research programs being carried out in various countries, and even these are being conducted within different agencies and different commercial companies within those various countries. Even the top trend lines are hard to detect!

Key questions are: (i) Who will develop the best technology for suppression of sonic booms? (ii) Will electronic aircraft win out due to environmental considerations and hypersonic services consequently be delayed? (iii) Will hypersonic aircraft services be of the Mach 2 to Mach 4 type systems largely being developed by agencies such as NASA, JAXA and ESA; or will they be more like the spaceplanes with Mach 6+ speeds that commercial organizations such as Virgin Galactic (i.e., SpaceShip2), Reaction Engines (i.e.,

Sabre), or Sierra Nevada (i.e., Dreamchaser) are developing [16].

The main scenario as to how hypersonic spaceplanes would operate is fairly well established. This would be that most spaceplanes would take off from key authorized airports and would climb at subsonic speeds to a quite high altitude of perhaps 16 kilometers (53,000 feet) before transitioning to hypersonic speeds and climb on a parabolic path that would peak at around 80 kilometers and certainly below 100 kilometers. On their descent to a level of perhaps 16 kilometers they would slow to subsonic speeds. They would deploy a telescopic long needle-like extension that would serve to create a series of a mini sonic booms rather than one huge sonic boom as they go subsonic. Finally, they would land at designated airports and cool down and then deplane passengers. Sierra Nevada's Dreamchaser, which is being developed to fly into orbit, as well as Reaction Engine's Sabre are, however, currently aimed at launching into space and would thus follow a much different flightpath.

Hypersonic Weapons and Defense Systems

The development of new systems for hypersonic flight, spaceplanes and new more efficient rocket systems has implications that go beyond the civilian applications discussed above. The future of NewSpace applications for military and defense applications will be discussed in Chapter 9, with most of this discussion being focused on the development of systems that would involve ballistic missiles flying in outer space. It is useful to know that there will likely also be significant uses of the protozone for defense-related purposes and that the research and testing of weapons systems that would fly below outer space are in active development.

In April 2018 the U. S. Department of Defense awarded a $928 million contract to Lockheed Martin to develop and test a new system known as a Hypersonic Conventional Strike Weapon. This weapons system would be operated on a 'boost-glide' basis, but defense research is continuing on air breathing rocket systems or scramjet technology as well. The U. S. Defense Advanced Research Projects Agency (DARPA) is seeking in its 2019 budget an increase of $148 million in spending on hypersonic research [17].

All of this research and weapons systems development in hypersonic vehicles is apparently being driven by increased spending by both Russia and China in these areas. Reportedly China has carried out twenty times more test flights of systems in this area and that the Chinese DF-17 hypersonic missile, with a range of 1,800 to 2,500 miles (2,880 to 4,000 km) has already been successfully test flown.

Meanwhile Boeing is under contract to develop a hypersonic Experimental Space Plane (i.e., the XS-1), and the company has come up with the more evocative name of the Phantom Express, which will also fly at hypersonic speeds. (See Fig. 6.7.) [17]

The current situation is thus that both hypersonic rocket boost and glide systems as well as spaceplane technology are being developed as weapons systems, with more money and more emphasis currently being placed on hypersonic rocket systems rather than hypersonic spaceplanes.

The current accelerated development of these various types of weapons could be used in a wide range of conventional

Fig. 6.7 U. S. Air Force Phantom Express spaceplane that could be used as a launch system. (Graphic courtesy of the U. S. Air Force.)

war-fighting capacities and could also be used to carry tactical nuclear weapons. The more that these types of hypersonic weapons systems are developed along with ballistic weapons systems, along with discussions of creating 'space forces' that will be charged with fighting war in space, the more endangered the concept of outer space being a global commons jointly available for the peaceful uses of these areas for all of humankind will become.

Conclusions

For many years there have been two prime areas that have been referred to since the beginning of the Space Age back in the late 1950s. These areas were outer space and air space. The International Civil Aviation Organization (ICAO) that was established by the Chicago Convention of 1948 was responsible for international coordination of air traffic management around the world. The United Nations also moved to create the Committee on the Peaceful Uses of Outer Space (COPUOS) that was key to the negotiation of the Outer Space Treaty of 1967 and the four subsidiary international agreements that came in the 1970s.

Today there is increasing recognition that there is a new area of concern and usage that lies between air space and outer space. This subspace, near-space or protozone has a growing number of potential new applications. These include uses for research and civil and defense needs as some regulatory oversight and safety coordinative activities will likely become more and more necessary. It is a sad reality that governmental action is seldom proactive to take preventive action before a serious problem or even a disastrous event takes place.

This chapter outlines the many new technologies and especially the growing number of applications that are now developing in this region above commercial air space (i.e., 20 kilometers or 12.5 miles) and below outer space, that is sometimes defined as 100, 110 or even 160 kilometers. There is more and more focus on the need to address space traffic management and control to avoid the ever-rising problem of orbital space debris. As the number of active and defunct spacecraft increases, especially with the expected near-term deployment of large scale small satellite constellations, and the amount of space junk climbs, the need for space traffic management continues to grow. As progress is made toward agreeing on space traffic management and control processes, it is equally important to recognize that this system needs to apply to not only Earth orbit but also to the subspace area that is referred to in this book as the protozone.

The trouble with many of the issues that one encounters in the area referred to as outer space is that these issues are often compartmentalized. This is to say that there are efforts to address orbital space debris, while others are concerned with such subjects as potentially hazardous asteroids, space weather, coronal mass ejections and solar radiation, and/or changes to Earth's magnetosphere. Yet others, such as the U. N. Committee on Disarmament Affairs, are concerned with weapons in space. Then there are the World Meteorological Organization (WMO) and the U. N. Environmental Program (UNEP) that are concerned with weather satellites, meteorological research, Earth observation and environmental effects.

There are many other sectors where U. N. specialized agencies and space-related matters intersect, such as food and agriculture, health, maritime related issues, etc. One of the goals of a recent book, entitled *Global Space Governance: An International Study,* that this author co-edited with Dr. Ram Jakhu of the McGill University Air and Space Law Institute, was to show how interconnected space policy and regulation is and the need to a holistic and more proactive approach to emerging space policy issues and concerns [18].

The UNISPACE + 50 meeting held in Vienna, Austria, in June 2018 also tried to relate space tools and system capabilities together with the U. N. 17 Sustainability Development Goals. This represents yet another effort to show the many ways space systems and uses interconnect with every aspect of human life in today's world. In efforts that seek to address the vital interconnections that exist between Outer Space usage and most aspects of human life, it should also be recognized that there is also a close connection between outer space and the protozone.

References

1. McDowell, J.: Where does outer space begin. Presentation to the manfred lachs conference, McGill University, Air and Space Law Institute, May 2017
2. Pappalardo, J.: New 'space fence' will spot space junk, small sats, and orbital weapons. Popular Mechanics, 16 April 2018. https://www.popularmechanics.com/space/satellites/a19831013/space-fence-update/

3. GESTRA – New space surveillance capabilities in Germany, 10 April 2018. https://idw-online.de/en/news692224
4. Peldszus, R.: Foresight methods for multi-lateral collaboration in space situational awareness (SSA) policy and operations. J. Space. Saf. Eng. **5**(2), 115–120 (2018)
5. Erwin, S.: Space news. In: Space Force: Pentagon Navigates Way Ahead and Awaits Direction Congress, 22 July 2018. https://www.space.com/40956-space-force-pentagon-awaits-congress-direction.html
6. Sheetz, M.: President Trump signs space junk directive aimed at cleaning up the cosmos, 18 June 2018. https://www.cnbc.com/2018/06/18/national-space-council-trump-signs-space-debris-directive.html
7. Wall, M.: For Sale: Balloon Rides to Near-Space for $75,000 a Seat. Space.com, 22 October 2018. https://www.space.com/23291-space-tourism-balloon-flights.html
8. Howell, E.: Space Tourists' Will Wear Special Socks on High-Altitude Balloon Flights. Space.com, 15 May 2017. https://www.space.com/36842-special-space-socks-high-altitude-tourists.html
9. Project Loon. https://x.company/loon/. Accessed 30 June 2018
10. WRC 2019 Agenda Item Details. https://www.transfinite.com/content/wrc2019list. Accessed 30 June 2018
11. Alleven, M.: Facebook ends ambitious Aquila program, will pursue partnerships instead. Wireless.com, 27 June 2018. https://www.fiercewireless.com/wireless/facebook-ends-ambitious-aquila-program-will-pursue-partnerships-instead
12. Athena satellite is now a facebook projecy. Space New Daily, 30 July 2018. http://www.satnews.com/story.php?number=447780911
13. Thessel, J.-P.: What's Up with Stratobus. https://www.thalesgroup.com/en/worldwide/space/news/whats-stratobus. Accessed 30 June 2018
14. Dark Sky Station. http://jpaerospace.com/. Accessed 30 June 2018
15. Pelton, J.N.: The international challenges of regulation of commercial space flight, Chapter 23. In: Pelton, J.N., Jakhu, R. (eds.) Space Safety Regulations and Standards, p. 299. Elsevier, New York (2010)
16. Pelton, J.N., Marshall, P.: Launching into Commercial Space, 2nd edn. AIAA, Reston (2015)
17. Button, K.: Hypersonic weapons race. Aerosp. Am. **56**, 21–27 (2018)
18. Jakhu, R., Pelton, J.N.: Global Space Governance: an International Study. Springer, Cham (2017)

7 On-Orbit Servicing, Active Debris Removal and Repurposing of Defunct Spacecraft

Introduction

There are many aspects of the NewSpace industries that tend to capture headlines. The idea of mining asteroids or new rocket systems that can be reused by landing them at precisely defined spots give rise to exciting television and helps to fire the human imagination. Two booster rockets landing together in synch produce great visuals that are immediately grasped even by small children. Not all of the NewSpace developments, like in-orbit refueling of a satellite, produce great theater.

The area of on-orbit servicing will appear to some to be the equivalent of taking a satellite into a service station for an oil change. At first glance on-orbit servicing might seem quite unexciting.

However, the truth is that this technology could revolutionize the economics of Space 2.0 industries. If we could cut the cost of space enterprises in a major way it could greatly expand the number of things we could do in space and make them profitable.

Some of the things that the new on-orbit servicing space companies are contemplating are straight out of *Star Wars*. We might learn ways to turn old spacecraft into shiny new satellites with many new capabilities. If we could recycle old satellites into new it might make a range of new space applications possible and also help clean up a lot of space junk in the bargain.

Today many of these activities are in the shadows. These shadows are created in part because a number of the activities are being led and funded by the Defense Advanced Research Projects Agency (DARPA) under their so-called Phoenix project. This thus tends to put such activities under the secrecy umbrella of national defense. Another shadow of sorts comes from the news media as a result of not covering activities that seem more mundane. The refueling of a rocket thrust system in space is just not as 'sexy' as a giant new big Falcon rocket boosting Elon Musk's Tesla sports car into space.

Yet these new on-orbit services may prove central to much of what happens in Earth orbit in years to come. Also, it

J. N. Pelton, *Space 2.0*, Astronautical Engineering,
https://doi.org/10.1007/978-3-030-15281-9_7

is interesting to note that organizations such as the DARPA are working with a range of start-up companies such as Altius Space Machines, NovaWurks and other lesser known names to invent this new technology [1].

New on-orbit servicing technology may prove key to removing space junk from orbit, manufacturing and processing of materials in space, building solar power satellites, recycling of defunct components and units in space by making new satellites out of old ones, and much more. The bottom line is that new technology developed by DARPA, NASA, other space agencies such as ESA, DLR, JAXA, the Chinese National Space Administration, and ROSCOSMOS, together with NewSpace industries, are now upping the ante on what can be done in space.

We are now seeing a raft of new ideas and capabilities as well as a redefining of the capabilities, dexterity, degree of autonomous artificial intelligence, as well as the cost efficiency of robotic systems in space. Progress in this area represents an unusual blend of sophisticated and high-cost space programs funded by civil and defense-related space agencies like NASA, ESA, DLR, JAXA and DARPA on one hand and Silicon Valley-like innovators from the Space 2.0 firms on the other. The key is not how big or well-established are the space innovators, but those who are able to come up with outside-the-box innovation.

What are today still largely governmental or defense-agency projects are increasingly giving rise to new commercial ventures in this area. Already there are three commercial ventures who are offering such on-orbit services as lifetime extension for commercial satellites, active space debris removal, or transfer of satellites from one orbit to another. These organizations are the U. S.-based ViviSat, the Canadian firm MacDonald Dettwiler Associates, who have developed a lifetime extension vehicle, and Cone-Express, which is a Netherlands-based firm that has designed their craft to uniquely fit into space available on the Ariane launch vehicle [2].

Sufficient progress has been made in this area so that efforts are now underway to create a new process for defining what might be called satellite servicing standards. This is not what might be called following a 'normal' international standards-making consultative process under the auspices of an international organization or association. Instead DARPA released a request for proposal for the Consortium for Execution of Rendezvous and Servicing Operations (CONFERS) Program in 2016, and on October 3, 2017, it selected a team consisting of the Secure World Foundation (SWF), the University of Southern California's Space Engineering Research Center (SERC), the Space Infrastructure Foundation (SIF), as well as Advanced Technology International (ATI) to serve as team leaders for this consortium [3].

Emerging New Capabilities in On-Orbit Serving Around the World

There are actually a wide range of new technologies and space-based capabilities that are under development at this time. These are systems to assist with or accomplish: (i) on-orbit servicing, refueling and component replacement; (ii) the capture and then repositioning of spacecraft to desired orbital positions; (iii) many different types of systems to assist with the active removal of space debris: and (iv) the development of robotic devices and artificially intelligent programs that are associated with refueling, installation of new components, sensors, antennas, and replacement batteries.

There are even more ambitious efforts to develop technologies, systems and artificial intelligence capabilities to convert or 'cannibalize' defunct satellites or orbital debris in order to make them into new usable spacecraft systems. Finally there are efforts, in many cases by the so-called space-mining companies, to develop new capabilities in terms both technology and new types of processing systems designed to allow space processing and space manufacturing, as well as to develop quality and materials standards for 3D or additive manufacturing in space.

The leader of many of these efforts is DARPA. DARPA has been engaged in R & D efforts in this area for over a decade. Projects have included: the Orbital Express mission in cooperation with NASA, the Spacecraft for the Unmanned Modification of Orbits (SUMO), and the Front End Robotic Enabling Near-Term Demonstration (FREND) mission. In its current on-orbit servicing vehicle project now underway, DARPA has involved a consortium of partners in an attempt to obtain greater commercial input on how to use the vehicle. This consortium consists of the Secure World Foundation, the University of Southern California's Space Engineering Research Center, the Space Infrastructure Foundation, as well as Advanced Technology International.

DARPA has established a revised set a mission objectives for its new test mission for on-orbit servicing that is scheduled for 2020 or 2021. These include working with an upgrading of an operational rather than a decommissioned satellite. They have also developed the ability to create a detailed 3D image of the satellite to facilitate a precision coupling with the satellite before using the FREND robotic arms grappling system to undertake system upgrades. Fig. 7.1 shows the FREND robotic arms grappling a large aperture communications antenna that is being transferred from a defunct satellite, to demonstrate how it might be repurposed in space.

Finally DARPA has broadened the scope to undertake the project with a commercial consortium as a partner. This is so as to allow the future development of a commercial on-orbit servicing organization that could work on governmental and commercial satellites and also seek to create international standards for on-orbit servicing. To date, a number of aerospace companies such as Airbus, MacDonald Dettwiler and Associates (MDA), Boeing, and Lockheed Martin have expressed interest in working with CONFERS in such a standards-developing effort [4].

As of November 2018 the members in in the CONFERS consortium had grown to 23, and on November 8, 2018, the first CONFERS conference was held in Washington, D. C., with a number of aerospace industries and governmental officials presenting their views on the future of the on-orbit servicing industry [5].

In addition, DARPA is pursuing a program known as Satlets, with new objectives as well. This is now considered a spin-off of the Phoenix on-orbit serving activities. Satlets are designed to

Fig. 7.1 Two FREND robotic arms simulating the manipulation of a satellite antenna in space. (Graphic courtesy of DARPA.)

be combined with components 'harvested' from defunct satellites. The current generation of Satlets are two-unit cubesat modules, each with a total mass of 7.5 kilograms. In their current incarnations these are known as eXCITE small satellites, or HI-Sats. Each satlet thus contains its own systems for propulsion, power, attitude control and memory. Space technology startup Novawurks is building the spacecraft using its Hyper-Integrated Satlet, or HI-Sat, product. This project can be scaled up from one module to hundreds of these Hi-Sats working as a constellation [6].

In addition to these independent development projects DARPA has worked with NASA on a number of efforts to develop robotic systems capable of demonstrating the ability to capture and couple with satellites and then carry out refueling and component replacement. This started with collaboration on the Orbital Express system in 2007 and then a collaborative test of the so-called Dextre system. (See Fig. 7.2).

Canada's Dextre robotic space helper was fabricated by MacDonald Dettwiler and Associates (MDA). This project was financed by the Canadian Space Agency. The Dextre system, also known as the Special Purpose Dexterous Manipulator, was specifically designed to work with and support NASA's Robotic Refueling Mission (RRM) experiment that was carried out on-board the International Space Station (ISS) in March 2012. This system is shown below in Fig. 7.3.

Fig. 7.2 Simulating the mating of the chaser Astro repair satellite about to capture the Next satellite. (Graphic courtesy of DARPA.)

Fig. 7.3 Dextre robotic system shown on the ISS. (Graphic courtesy of NASA.)

The Dextre system, however, can be used in space for other purposes that include external maintenance activities for the ISS, support for astronaut extravehicular activities – especially to carry out equipment maintenance and fine motion actions related to calibrations – retrofitting of equipment, or in performing tests or experiments. These capabilities could even be used in the future to support space processing and manufacturing [7].

Dextre is one of the reasons why MDA is considered one of the leading commercial firms to develop commercially based services in the area of active debris removal, repositioning of satellites to new orbits, and in-space processing and manufacturing.

The Dextre system was designed as the "hands" for the Space Station Remote Manipulator System (SSRMS). Thus Dextre is able to give the ISS robotic Mobile Servicing System (MSS) the ability to perform fine dexterous tasks. The satellite was launched to the ISS on STS-123 in March 2008 and is controlled entirely from the ground. The hope is that this type of proven capability can be useful for a wide range of future space operations that include on-orbit servicing, on-orbit refueling and perhaps ultimately space processing and manufacturing operations. This type of capability, of course, can also be of use in active debris removal operations as well.

Coping with Space Debris

The issue of space debris has only grown in importance in the past two decades. This problem was greatly increased in 2007 by the Chinese launching of a missile to purposely hit a defunct weather satellite. The explosion created over 2,000 new debris elements over 10 centimeters in size. This was followed by the 2009 accidental collision of a defunct Russian Cosmos satellite with an Iridium mobile communications satellite that also created over 2,000 new debris elements (Fig. 7.4).

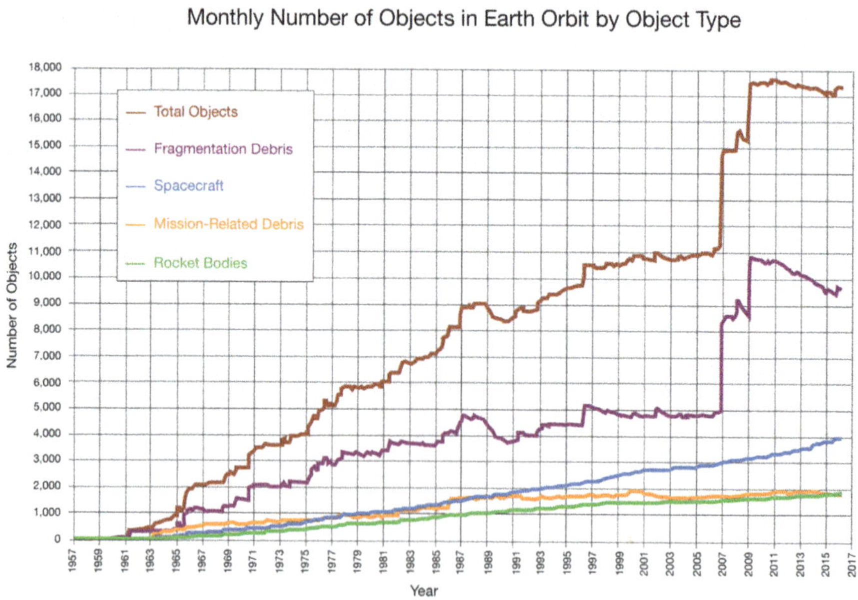

Fig. 7.4 Chart showing the steady rise in space debris and sharp increases in 2007 and 2009. (Graphic courtesy of NASA.)

There is urgent interest in seeking to remove the largest defunct satellites in low Earth orbit, such as Envirsat, which is over 10,000 kilograms in size. In addition, the United States has just issued a new Space Directive-3 that is aimed at providing for improved space situational awareness and improved procedures for space traffic management. Scott Pace, Executive Secretary of the National Space Council, has described this process as follows: "The reforms, which will be enacted over the coming months and years, will be specific to the United States rather than negotiated through the United Nations." Pace has explained that "the space council opted for a bottom-up process in the name of expediency rather than trying to create an international treaty. By setting a proper example, the United States intends to establish norms that Europe, China, Russia, and others working in space will follow" [8].

Engaging in Active Debris Removal

There remain a number of issues to be resolved as to the regulatory environment that would allow governmental or private entities to engage in active debris removal. Some believe that an amendment is needed to the Liability Agreement that places responsibility for space objects with the registered launching state rather than with the actual owner and operator of a spacecraft. Currently it is only the launching state that is held responsible for all generated space debris. The key point is whether liability for space objects might be transferred from one entity to another and if "absolute liability" might be transferred from one entity to another for debris striking Earth.

Some of the latest significant developments are in the areas of licensing of satellite launches, updating of space oversight processes, and space traffic management. These subjects have all been addressed in the U. S. Space Policy Directives issued by the United States in April and June of 2018. (See Appendix D at the end this book for the full text of these directives.)

The main thrust of these latest U. S. space policy directives has been to say in effect that there would be a national effort to increase the speed of licensing of satellite systems, particularly large-scale constellations, on one hand, but also to improve the accuracy of space situational awareness and upgrade procedures in space traffic management by the United States on the other hand. The hope is that other spacefaring nations would follow suit.

While the regulatory and global space governance aspects of this sensitive matter continue to be considered in international forums such as the U. N. Committee on the Peaceful Uses of Outer Space (COPUOS) and in its Working Group on the Long-Term Sustainability of Outer Space Activities (LTSOSA), actual efforts to improve both space situational awareness and space traffic management processes continue.

There are now a number of commercial entities that are tracking space policies around the world, and the U. S. Space Policy Directive 3 directs the U. S. Space Command to take these sources of information into account in addition to using the additional resources of the new S-band radar addressed earlier in this chapter [9].

In addition there are a number of Earth-based laser and directed energy systems that might be employed to redirect the orbits of space objects that are in danger of potentially colliding with other space objects that would create more space debris. These types of systems are probably the most cost-effective way to avoid space object collisions in the shorter term, but there is a wide consensus that this is not likely to provide a long-term solution. Either governments or commercial operators must find that. There have even been suggestions that a form of space insurance process might be devised to implement a commercial-type approach to the space debris problem over the longer term [10, 11].

Meanwhile there are a number of efforts that are going forward to develop the actual technology that might be able to address the orbital debris effort. Many of these efforts involve such ideas as having passive systems deploy at the end of life that have little cost. This might be something such as balloon systems to create atmospheric drag in order to bring low Earth orbit satellites down faster. There are other ideas that would involve sending up satellites that could spray out epoxy materials or harpoon a defunct satellite with netting materials. Again the idea would be to create atmospheric drag to bring down a satellite so it would burn up in the atmosphere. There are now dozens of "active" deorbit technologies, as well as less costly passive systems, based on increased atmospheric drag, now under consideration. The current plans to launch many thousands of new satellites now make these new plans to de-orbit existing satellites much more urgent.

The following, however, represent some of the current programs that are seeking to develop and operate systems that could carry out active debris removal. A number of these efforts represent commercial initiatives [12].

ConeXpress Orbital Life Extension Vehicle (Sometimes Called ConeX or CX-OLEV)

This initiative of Dutch Space of Leiden is based in the Netherlands, but it is closely associated with ESA and Arianespace. The initial concept that gave rise to ConeX was to use what is called "the standard Ariane 5 conical payload adapter" as the main feature of its design. This project is seeking to create a propulsive unit that would fit in as a flattened cone-shaped mechanism uniquely designed to fit as an integral part of the Ariane 5 launch vehicle. It would have to be readapted to the Ariane 6.

The estimated cost of a ConeX mission has been projected to be about 35 million Euros. The system could be used to lift a spacecraft outward from LEO to GEO in a spiral orbit, or to deorbit a defunct space object. The ConeX would be powered by an electrical propulsion system that derives from the design developed by ESA for the Smart-1 lunar mission. Currently there are no active customers to use such a device for either rescuing a spacecraft that failed to reach GEO orbit or for actively removing a defunct satellite in a dangerous orbit. (See Fig. 7.5.)

Fig. 7.5 Artist's impression of ConeXpress Orbital Life Extension Vehicle operating in space. (Graphic courtesy of ESA.)

MacDonald Dettwiler and Associates and Space Infrastructure Servicing

McDonald Dettwiler and Associates (MDA) has long been a leader in developing space robotic systems. They have also been one of the first commercial firms to develop systems that could be used as free flying on-orbit servicing vehicles due to their work for NASA and DARPA. They have developed what they call a Space Infrastructure Servicing vehicle. This SES vehicle is a multi-functional unit designed for on-orbit servicing, refueling or retrofit, for repositioning a satellite in the 'wrong' orbit, or as a means for active debris removal. In early 2011 MDA had arranged with INTELSAT to undertake a servicing mission, but the final contractual arrangements were never concluded, and the agreement was thus canceled in 2012. MDA and its subsidiary SSL is currently seeking to find specific commercial or governmental customers to actually commission in-orbit missions. It has since altered its approach to organize its on-orbit servicing offerings into a partnership that includes MDA, Draper Laboratories, SSL, DARPA and the Naval Research Laboratories [13].

The Satellite Infrastructure Services (SIS) website indicates that SIS will be able to offer on-orbit services on a contractual basis within the next two years. Its promotional offering, for this type of on-orbit service, states the following: "Space Infrastructure Services, SIS, offers the world's first on-demand

Fig. 7.6 The Vivasat MEV shown as mated with a satellite for servicing purposes. (Graphic courtesy of Vivasat.)

robotic service spacecraft available for missions in 2021 and beyond. These can be pre-scheduled or as an emergency call-up servicing – like roadside assistance in space. Services are insured and payment is not due until successful service completion, and your satellite continues to operate during most SIS on-orbit robotic servicing procedures. Our servicer is compatible with government and commercial spacecraft – even those not designed to be serviced in space." These services could of course be used for active space debris removal, but the economic costs that would likely be above $40 million suggest that these services would be for lifetime extension or orbital rescue of a quite expensive spacecraft [14].

Vivisat and Its Mission Extension Vehicle

Vivisat, which is U. S.-based, has developed what it calls its Mission Extension Vehicle (MEV). Vivasat was organized as a direct competitor to SIS. This MEV on-orbit service has been presented as a means for refueling and repairing satellites, but it could also be used for repositioning of spacecraft or to undertake active debris removal. It is advertised as being very flexible, in that the MEV was designed to mate with virtually all of the roughly 500 geosynchronous application satellites currently in orbit or now scheduled for launch. The SIS vehicle has now also been redesigned so that it can capture and mate with spacecraft not designed for on-orbit servicing. (See Vivasat Mission Extension Vehicle in Fig. 7.6.)

The truth is that there has not been any truly demonstrated commercial market for on-orbit servicing of satellites or for active debris removal. The significant amount of effort and resources that the DARPA has devoted to this type of effort through its Phoenix and other programs could change the attitude toward on-orbit servicing going forward.

X-37B OTV – NASA, U. S. Air Force and DARPA

Some have indicated that other initiatives, such as the U. S. Air Force X37B OTV, might be used for reclaiming and re-deploying space resources in a cost-effective manner as a reusable vehicle. The X-37B orbital test vehicle has been developed as an experimental, reusable spaceplane. It is somewhat like a small shuttle and is unmanned and completely robotic. This is now a classified project after this project was turned over to the U. S. Air Force from NASA. It is thus only speculation to suggest that such a vehicle might be used as a mechanism for repairing or refueling malfunctioning satellites, or for returning classified spacecraft back to Earth.

Sierra Nevada Dreamchaser Spaceplane

Yet another alternative for a vehicle that might be used for on-orbit servicing or to deploy systems to initiative de-orbit of large defunct objects in low Earth orbit is the Sierra Nevada Space Plane that has now been dropped from the competition for a means to fly astronauts to the International Space Station and provide a return capability.

The Remove Debris Small Satellite

On June 20, 2018, the Remove Debris proof of concept small satellite was launched by Nanorocks from the Kaber launch facility on the ISS. This small satellite with a mass of some 100 kilograms was constructed by Surrey Space Technology, Ltd. after some five years of development [15].

This experimental small satellite is designed to test the feasibility of various concepts that have been proposed for lower cost ways to create passive systems that might accelerate the deorbit defunct space object. The small satellite consists of the following component parts: (i) two cubesats; (ii) a net and a harpoon; (iii) a laser ranging instrument; and (iv) a "dragsail" designed to unfurl behind the main satellite. The idea is to test these various components to create significant drag to help space junk develop aerodynamic resistance to Earth's atmosphere and thus decay. This is only a test project of techniques, but if these tests are deemed successful it could form the basis of other actual projects to use the techniques in the future with actual debris objects [16]. (See Fig. 7.7.)

German DLR DEOS Mission

Another effort to demonstrate active debris removal is the so-called DEOS mission that is being carried out by the German space agency DLR. This project, which has a target satellite and a chaser satellite that will capture the target satellite and demonstrate its removal from orbit, is, in many ways, akin to the DARPA/NASA Orbital Express mission completed in 2007. It has precise sensors that will ensure that the capture in orbit is successful and that the mating is accomplished without damaging to the chasing or target satellite.

CleanSpace One – EPFL

A miniature version of the Orbital Express and the DEOS mission is the experimental test known as CleanSpace One. This mission is being carried out by a team of organizations in Switzerland. The project, which is using

Fig. 7.7 The Remove Debris small satellite being readied for release by the Nanoracks Kaber launch system via the JAXA Kibo airlock on the ISS. (Graphic courtesy of NASA and Nanoracks.)

only cubesat technology, is being led by the École Polytechnique Fédérale de Lausanne (EPFL) and The Swiss Space Centre – a division of EPFL. Another partner was the Swiss Space Systems (S-3) that has now declared bankruptcy, and this may delay the project.

EDDE – Electro-Dynamic Debris Eliminator

Currently there are three mainline strategies to diminish orbital space debris. These three concepts are: (i) the use of land-based energy transmission systems that could over time help divert orbits of objects projected to collide; (ii) the use of passive systems to speed deorbit due to the creation of aerodynamic drag; and (iii) active debris elimination by systems that could drag debris down swiftly. There is, however, yet one other concept of particular interest. This is the proposed deployment of a space system that could be sustained in space by using Earth's magnetic field to create an electronic propulsive force to deploy a device of 'passive nets' to create atmospheric drag to deorbit debris. NASA in 2012 awarded a $1.9 million development contract to the Star Technology and Research (STAR) Company. This system, as shown below in Fig. 7.8, is known as the Electro-Dynamic Debris Eliminator (EDDE) [17].

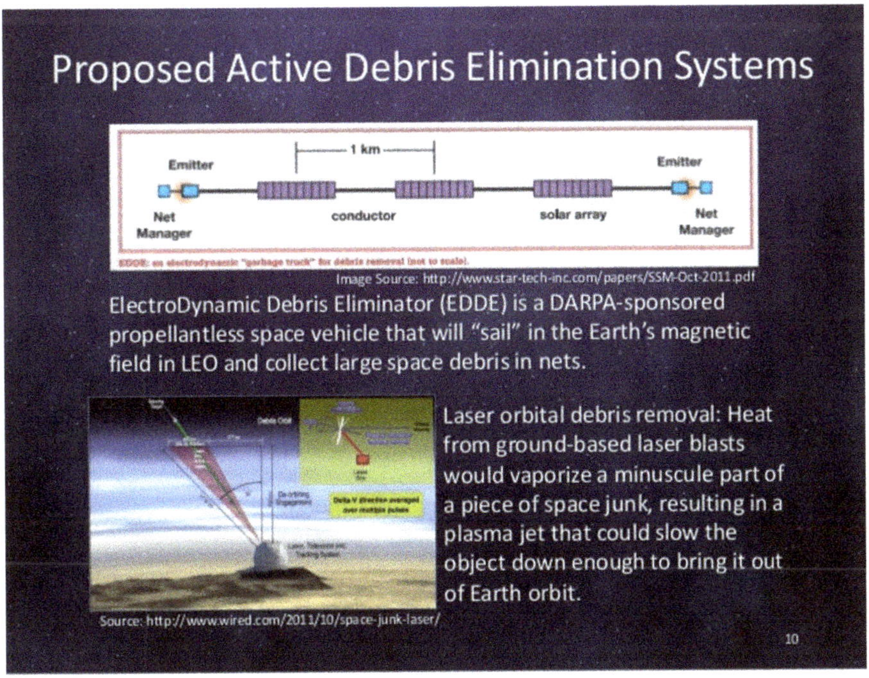

Fig. 7.8 Conceptual designs for active debris removal such as an Electro-Dynamic Debris Eliminator (EDDE) or Laser Heat System from the ground.

Conclusions

Some of the biggest changes and innovations that seem to be coming in the field of satellite applications are now being driven by DARPA. This applies to such activities as the development of spaceplane systems and various aspects of the so-called "Phoenix" projects that include sophisticated robotic capabilities in space, refurbishment and retrofit of satellites in orbit, the possible cannibalization of defunct satellites to redeploy parts from these satellites into a new system, and new ideas about satlet units for many flexible new uses. There are also new capabilities originating here related to better space situational awareness, space traffic management, active debris removal, and improved command and control capabilities related to space operations. It does not seem to be an exaggeration to say that all of these technical and process innovations will likely lead to new commercial opportunities that will follow from the DARPA-driven efforts to truly transform the field of space applications and to advance the cause of new Space 2.0 industry innovation.

There are changes that are driven by Space Directive 2 and Space Directive 3 that are seeking to improve space situational awareness processes and improved use of private capabilities to track space debris, space objects and satellites in orbit. These directives are aimed at increasing capabilities to carry out space traffic management and shifting regulatory oversight responsibilities to the new Office of Space Commercialization in the Commerce department.

The bottom line is that two major trends are coming together. One trend is concern about orbital space debris, the need for improved space situational awareness, and finding better ways to engage in space traffic management and reduce debris objectives. The other trend is the development of new space technologies and the burgeoning of new more cost effective capabilities that are coming from the world of Space 2.0 and new ways of doing things in space. These two trends are spurring innovations in the world of on-orbit servicing and in new ways to cope with orbital debris. Research from the world of defense and space agencies and innovation from the world of Space 2.0 are coming together to create new capabilities and new opportunities.

Meanwhile the volume of space debris continues to grow. The latest studies from ESA project a major collision occurring approximately once every five years. Low Earth orbit and Sun-synchronous polar orbits are of particular concern. Today tracking systems are monitoring and charting the orbits of approximately 22,000 objects in low to medium Earth orbit that are at least 10 centimeters or larger (or about the size of a softball). Perhaps of even greater concern is that there are now some 700,000 objects 1 centimeter or larger in size that are creating a worrisome and constant avalanche of space debris that increasingly blankets low Earth orbit at a time when many thousands of satellites in LEO constellations are proposed to be launched within the next few years. Clearly this is both of concern and an opportunity to innovate in the field of on-orbit servicing and debris-removal activities.

References

1. Leone, D.: DARPA Selects contractors for phoenix satellite servicing program. Space News. http://spacenews.com/darpa-selects-contractors-phoenix-satellite-servicing-program/ (2012)
2. Pelton, J.N.: New Solutions to the Orbital Debris Problem, Chapter 2. Springer, Basel (2014)
3. Industry Partners Help Establish Consortium for Satellite Servicing Standards. https://mail.aol.com/webmail-std/en-us/suite (2018)
4. Caleb, H.: DARPA revamps Phoenix in-orbit servicing program. Via Satellite. https://www.satellitetoday.com/government-military/2015/06/02/darpa-revamps-phoenix-in-orbit-servicing-program/ (2015)
5. Global Satellite Servicing Forum. https://mail.aol.com/webmail-std/en-us/suite (2018)
6. Caleb, H.: DARPA trying to launch Smallsat experiment on an Indian rocket. Space News. http://spacenews.com/darpa-trying-to-launch-smallsat-experiment-on-an-indian-rocket/ (2017)
7. Harding, P.: Dextre and RRM Complete Record Breaking Week of Robotics on ISS, NASA Spaceflight. https://www.nasaspaceflight.com/2012/03/dextre-rrm-complete-record-breaking-week-robotics-iss/ (2018)
8. Berger, E.: As mega-constellations loom, US seeks to manage space debris problem. ArsTechnica. https://arstechnica.com/science/2018/06/as-space-gets-more-crowded-us-seeks-to-ensure-a-safe-environment/ (2018)
9. U.S. White House: Space policy directive-3, National Space Traffic Management Policy. https://www.whitehouse.gov/presidential-actions/space-policy-directive-3-national-space-traffic-management-policy/ (2018)
10. Pelton, J.N.: New Solutions to Orbital Debris Problems. Springer, New York (2015)
11. Pelton, J.N.: Possible institutional and financial arrangements for active removal

of orbital space debris. In: Pelton, J.N., Allahdadi, F. (eds.) Handbook of Cosmic Hazards and Planetary Defense. Springer, New York (2015)

12. ConeXpress-Orbital Life Extension Vehicle (CX-OLEV). https://artes.esa.int/projects/conexpress-orbital-life-extension-vehicle-cx-olev (2018). Accessed 3 July 2018
13. SSL: On-orbit servicing changes everything. http://www.spaceinfrastructureservices.com/ (2018). Accessed 3 July 2018
14. Ibid
15. "NanoRacks Deploys Largest Satellite from International Space Station to Date". https://mail.aol.com/webmail-std/en-us/suite (2018)
16. Clark, S.: Space junk clean-up demonstrator deployed from space station. Spaceflight. https://spaceflightnow.com/2018/06/26/space-junk-clean-up-demonstrator-deployed-from-space-station/ (2018)
17. Azriel, M.: Electro-dynamic debris eliminator. Debris Eliminator. http://www.spacesafetymagazine.com/author/merryl-azriel/ (2012)

Space-Based Solar Power Satellite Systems

8

Introduction

One of the unrealized potential uses of space systems that has been discussed and examined for nearly five decades is the tantalizing idea of creating solar power satellite, or what is most commonly now called space-based solar power (SBSP). The theory is that it would be possible to create such a system at geosynchronous orbit or perhaps at a suitable Lagrangian point that would be capable of beaming clean energy back to Earth on a 24 hours a day 365 days a week basis. This energy would be sent down to Earth using a suitable radio frequency or laser transmissions that could then be converted to electrical power using diodes placed within dipole antenna receivers. This energy could then be sent to locations where energy is most needed. Countries without large energy reserves, such as China and Japan, have accordingly been among those nations that have pursued some of the most active research programs in this area [1].

The economic importance of satellite telecommunications, broadcasting, remote sensing and space navigation is now well established. The satellite communications market, which represents about $150 billion a year in revenues, is today, however, the only true economic powerhouse as space industries go.

Some believe we might well be on the verge of creating in future years yet another new space industrial sector. This new space application has the potential to become another truly huge new space industry market. Clean and green electrical power from space is thus one of the aspirations of those seeking new commercial space applications.

In 1972 on the Voyage beyond Apollo this author had the pleasure to meet Dr. Peter Glaser, the father of what was then called solar power satellites, or Sunsats. Glaser advocated creating space systems that displayed a large number of photovoltaic cells that could be illuminated by the Sun not just during the day but on a continuous basis. Glaser advocated that this power be transmitted back to Earth in the microwave bands. One of the many highlights of this amazing cruise was to hear Dr. Glaser debate with English engineers whether RF

J. N. Pelton, *Space 2.0*, Astronautical Engineering,
https://doi.org/10.1007/978-3-030-15281-9_8

transmission, or as the English scientists argued, perhaps lasers might be the best way to proceed. In the early 1970s petroleum was cheap. The world population was only 4.5 billion. At this time in history neither carbon-based fuels nor climate change were widely seen as looming issues. Thus solar power satellites were not seen as a practical nor cost-effective way to derive electrical energy, but rather something for the possible longer term future.

In only half a century the world has changed greatly. Earth's population has swelled to 7.5 billion and may expand to as much as 12 billion by 2100. Climate change and global warming are now major environmental concerns worldwide. It is possible to envision SBSP system designs that will, in the not too distant future, perhaps, be cost-competitive with carbon-based energy systems such as coal, oil, or nuclear power. Despite some health and environmental concerns, SBSP systems are largely seen as a "green" energy source that could be distributed to parts of the world where significant amounts of clean energy are needed.

NASA scientist Geoffrey Landis, who has spent years researching the technical and economic feasibility of SBSP systems, contributed an important idea to the field. He noted that one of the key challenges in many cities was not continuous supply of energy on a 24 hours a day basis but simply meeting peak-load requirements that are typically only about one-eighth of the total load requirement. Thus one might consider a role for SBSP systems would be to supply peak load requirements from a GEO orbit location to different cities as they encountered peak requirements. A phased-array or spot beam system could provide peak-load requirements to different cities on demand [2].

There are now safety standards for power beam density for SBSP system transmissions, and there are health standards for what level of power might be received by land-based or sea-based rectennas that are considered safe for receiving the transmitted energy. Most experts agree that RF transmission is safer than lasers and would have less of a problem of penetrating cloud cover.

There has been sufficient progress in the area of SBSP systems that a variety of national projects and even commercial ventures have begun to come forward to suggest that they could in coming years successfully deploy SBSP systems that would be cost competitive and would pose no health risks if designed and operated to current safety and health standards.

There have been several starts and stops by commercial ventures that have looked promising but have faltered for a variety of business, economic and technical reasons. At the outset the idea involved what was seen as a single giant facility that included massive solar arrays, the system to convert the power to be transmitted and the transmission system back to Earth and receiving system on the ground. Today SBSP systems are more frequently seen as coming in several different parts. These parts include low mass concentrators that focus the Sun's radiation so that a smaller unit of photovoltaic cells (or quantum dot) solar power generators would see the equivalent of many suns. In space there would be then be three key elements: (i) the solar concentrators; (ii) the photovoltaic power generators; and (iii) the unit that would convert the energy to a microwave or millimeter wave transmitter that would beam the power down to Earth. On Earth there might be several large-scale antenna systems (called rectennas) that could receive the radio frequency (RF) power at different locations and convert it to electrical energy at much lower frequency and transmit that power to urban transformers for distribution.

The logic of these systems and key economic calculations have for a long time focused on the ability of space-based systems to become cost-competitive with conventional electrical energy plants that are fueled by oil, coal, or nuclear-based power. What is important to note is that today this is only a part of the cost-analysis equation.

Tremendous progress is now being made with new types of Earth-based renewable energy sources. These increasingly cost-efficient electrical power sources include, among a growing number of options, solar (photovoltaic systems and soon quantum dot systems), hydroelectric, geothermal, wind farms, and even passive energy conservation systems such as better insulation and lighting systems. These ground-based renewable energy systems are all increasingly cost effective and are now developing apace. All of these renewable systems are not only becoming more cost-effective but are serving to reduce reliance on a consolidated energy grid that has positive implications in terms of resiliency. These are all key factors to consider with regard to the potential future of SBSP systems. Further when one considers the cost of coal, oil or nuclear-based energy systems, it is important to include the environmental costs associated with these, rather than conveniently ignoring them.

This chapter explains the current status of the most advanced solar power satellite system designs in space and for ground-based rectennas as well as the financial, economic, management and regulatory issues that still need to be overcome to move solar power satellites from engineering prototypes to actual operating systems.

From Early Design Concepts to Current SBSP Systems

The first concepts with regard to the possible development of solar power satellites came at the same time that there were also thoughts of possibly designing and manufacturing very large communications satellite platforms. In the 1970s and 1980s there were also very futuristic ideas of being able to carry out materials processing and manufacturing on the Moon. These included ideas that satellite antenna reflectors and solar arrays might thus be manufactured on the Moon. This would allow the communications platforms and solar power satellites to be built on the Moon and then "lowered" to GEO orbit with only the intricate electronics being manufactured on the ground and then added to these large and massive structures.

The logic of much of this thinking was that the launching costs for very large structures had remained quite high and seemed likely to remain so for many years to come. Launching materials or structures off the Moon with its modest gravity could open up major new opportunities.

The enthusiasm that came with the sending of astronauts to the Moon starting in 1969 gave rise to unwarranted enthusiasm as to what might be possible next. The president of Caltech predicted sending astronauts to Mars in the 1980s. The sky was no longer the limit. Instead, though, the United States and much of the spacefaring world restricted its vision in the decades that followed to projects like the space shuttle, low Earth orbit space stations, and modest progress in the field of space.

Only in the last few years has enthusiasm for space and the "we-can-do-everything-that-we-dream-of" mentality begun to return to space enterprise. In the preceding chapters, the remarkable new achievements in satellite communications, remote sensing, space navigation, meteorological satellites, and on-orbit servicing have certainly been aided by lower cost launch vehicles. Further the prospect of reusable launch vehicles and new launch options for cubesat-sized spacecraft have helped to foster the rise to Space 2.0 initiatives and entrepreneurial outside-the-box thinking. But all of these initiatives were able to proceed without innovations in launch vehicles. Indeed some of the small sat innovations may have even been aided by higher launch costs. This is to say that of all of the space applications discussed up to this point space-based solar power is uniquely in need of significant cost reductions in launch costs. In essence, better and less costly launchers are essential to achieving financial viability for Sunsats.

Current Efforts to Design and Develop Space-Based Solar Power

There are many national space agencies, research organizations and commercial space organizations that are pursuing new and viable designs for space-based solar power systems and various types of technology that could contribute to the development of these. The various technologies that are under consideration or development include the following:

- solar concentrators with lower mass with increased reflectivity
- protective systems to preserve the performance of photovoltaic cells
- higher performance photovoltaic cells such as those operating in the ultraviolet range
- quantum dot technology to improved power generation capabilities
- material processing in space to generate refined materials to make solar cells
- additive manufacturing techniques to create solar cell arrays in space
- improved, lower cost and reusable launcher systems
- improved diodes and dipole RF antennas for rectennas
- means to reduce interference between SBSP systems and communications systems
- improved robotics for the deployment and repair of SBSP systems
- phased-array systems to allow distribution of RF signals to rectennas more efficiently

There are today a wide variety of concepts that have been proposed, such as the Alpha SBSP systems. This design was developed by Dr. John Mankins of Mankins Space Technology, Inc., who led the NASA research effort in this area for a number of years. (See Fig. 8.1.)

Another concept approach comes from a NASA design study that tends to validate the three parts approach of: (i) a modular set of low mass yet large aperture solar concentrators; (ii) photovoltaic cell power generators; and

Fig. 8.1 The Alpha space-based solar power system. (Graphic courtesy of John Mankins of Mankins Space Technology, Inc.)

(iii) RF transmitters. These concentrators could allow the solar cell power generation system to be able to be subjected to the equivalent of seeing many dozens of suns. (See Fig. 8.2.)

A recently released design from Space Energy, Inc., a commercial organization, claims it will deploy a space-based solar power system that in many ways has similarities to the NASA design in terms of having being modular [3]. (See Fig. 8.3.)

Peter Sage, the CEO of Space Energy, Inc, has expressed optimism that their system can be fully deployed and operational by 2025. He has been quoted as saying: "This is an inevitable technology; it's going to happen. If we can put solar panels in space where the Sun shines 24 hours a day, if we have a safe way of transmitting the energy to Earth and broadcasting it anywhere, that is a serious game changer." The problem is other commercial organizations that have gone before and claimed that they could sell space-based derived solar power at competitive rates but have not been able to deliver on their aspirational statements.

The great challenge for these systems is not only to be competitive with oil, coal and nuclear-based electrical power systems but renewable energy sources as well. The costs of terrestrial solar cell units and wind turbine electrical generators, for instance, have fallen sharply in the last decade. Solar cell systems, in particular, have fallen in price by a factor of 77% and are expected to drop a similar amount in the future as the cost of

Fig. 8.2 NASA concept for a space-based solar power (SBSP) system. (Graphic courtesy of NASA.)

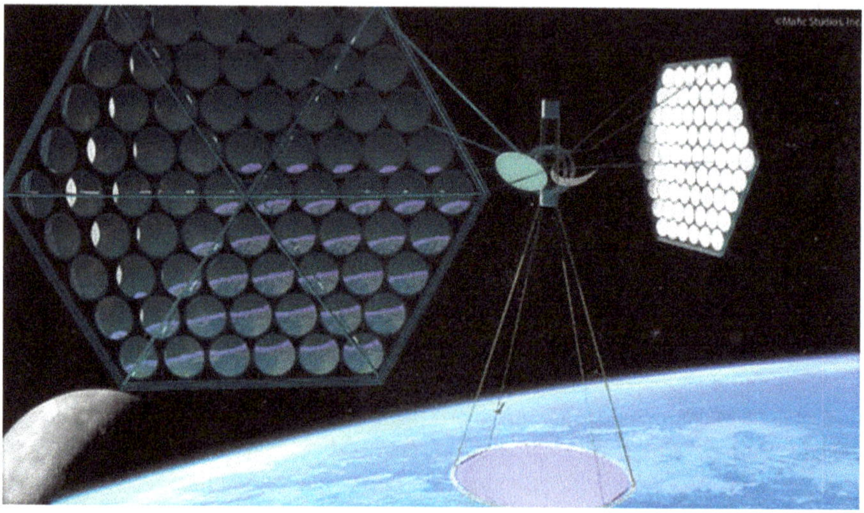

Fig. 8.3 The proposed design by Space Energy, Inc. (Graphic courtesy of Space Energy, Inc.)

energy storage systems also continues to fall. This seems to be especially the case with the Chinese manufacturers entering this highly competitive market for both solar cell power systems and more cost-effective energy storage systems [4].

Ground Systems for Space-Based Solar Power Reception

The key to the operation of SBSP systems on Earth will be in the design and operation of what are called rectennas. This is actually short for "rectifying antennas." It is usually envisioned as a large series of dipole antennae configured as a network over a large circular area thousands of square meters in size.

Each of the dipole antennas would be configured with a radio frequency (RF) diode connected across the dipole elements so that it would 'rectify' the alternating current (AC) to induce a direct current (DC) power flow. This power flow from each diode could then be consolidated and transmitted to a transformer within an electric grid network for distribution to users.

The rectenna would spread out over a large area that would include thousands of such dipoles, each with diodes arranged so that they would convert on a consistent basis the incoming RF transmission into DC power for the entire area. The RF transmitting antennas out in space could be configured as spot beams. This would allow the sunsat to distribute its transmissions to different but quite specific locations that were in proximity to where the power was actually needed. The reason that the power would be transmitted in this manner is to avoid concentrating it and creating a form of death ray. A very high intensity beam would be dangerous to anyone in an aircraft or helicopter flying through the beam. In addition to health concerns about the radiation levels being too concentrated, there is also the additional concern that there could also be reflected energy returning power back up to space that would create harmful interference to satellites in orbit.

The invention of diode-based rectennas is attributed to William Brown, an American engineer that came up with the idea in the mid-1960s and was issued a patent in 1969. Rectenna technology can be used in the context of wireless power transmission. Other potential applications could include maintaining platforms or even keep a helicopter or other type of aircraft aloft. The most important application in terms of space systems would be to develop rectannas to receive RF transmissions from space-based solar power systems.

Research in this area is currently focused on the ability to create so-called devices that would operate not at the RF level but in the nanometer frequency range of infrared and light waves. Ultrafast and miniaturized diodes that are scaled to operate in these frequencies could, in theory, also be used to convert light directly into electricity. This type of device is called an optical rectenna, or nantenna. High efficiencies up to 70% conversion might be attained, in theory, but efficiency has so far been limited, even though working optical rectennas have now been demonstrated [5] (Fig. 8.4).

International Initiatives in the Space-Based Solar Power Arena

SBSP is being actively pursued by Japan, China and India in particular, but there are also certainly significant and ongoing efforts in the United States, in Europe and Russia, among others. There are several international efforts to share

Fig. 8.4 Illustration showing a rectenna for SBSP reception plus an electrical power transmission system to an urban center.

information with regard to the development of new technology in this area. Perhaps most notable is the ongoing effort to share research and development in the SBSP field among the following organizations: the University of Science and Technology Beijing, the China Academy of Science and Technology (CAST), the Indian Space Research Organization (ISRO), the Indian Institute of Space Science and Technology (IISST), the Japanese Aerospace eXploration Agency (JAXA), several Japanese universities, and others.

In 2008 Japan's Diet passed its Basic Space Law. This law established space solar power as a national goal. As a response to this law the Japanese space agency (JAXA) has developed a roadmap, or plan, for its ongoing efforts in the development of SBSP technology and systems. It has also encouraged commercial SBSP ventures. JAXA is to some extent unique in that it has developed goal-oriented programs built on applications and uses defined by its national legislature in its Basic Space Law. Thus JAXA has developed roadmaps and even timetables in such areas as national and international broadband services, remote-sensing applications, and now sunsats as goals to achieve. JAXA has designed its ongoing technological and applications programs to coincide with these national social and economic objectives. This has tended to make JAXA more applications and uses driven rather than technology driven. All space agencies have strategic plans, but JAXA roadmaps are more driven by clearly defined national economic and social goals than the other agencies [6].

In 2015 the China Academy for Space Technology (CAST) showcased their roadmap to a 1 GW commercial system at the International Space Development Conference (ISDC).

India has carried out extensive reviews of their current energy production capabilities and what their projected increased energy needs will be, especially in the case of sustained economic growth rates in the range of 5%

per annum up to 8% per annum. Even a sustained growth rate averaging 7% per annum would result in the need for nearly 1,400 gigawatts capacity by 2050. This would require an increase in capacity of more than a hundred times India's current capacity. India has committed to greatly increasing its domestic energy production based on terrestrial renewable energy sources, but has sought to work with other nations to explore the possible development of SBSP systems as a supplementary source of energy.

The result of India's study was to conclude that the pathway to finding an economically viable way to deploy any truly meaningful sunsat capacity would require the development of reusable launch system that could lift needed payloads to orbit with an efficiency of $100 to $200 per kilogram. This also assumed a reusability factor of 100 to 150 times for the launch systems. In short, India and ISRO in particular proposed that an international consortium would need to commit to such a development enterprise and that India's prime contribution would be in the development of new launch systems that would be required to deploy the needed photo-voltaic (PV) cells to orbit. Currently the cooperative agreements to work on new sunsat systems is limited to China, Japan and India, and the needed breakthroughs in the launch systems and the sunsat technology have not come close to being attained [7].

Other Reasons for the Development of SBSP

There are actually several reasons why there seems to be continuing interest in the development of space-based solar energy. The first and foremost reason is the favorable environmental effect. It has been claimed that sunsats are a clean and reliable way to green and renewable energy that is reliably available 24 hours a day, 365 days a year. Countries such as China and Japan, for instance, have projected their future energy growth needs and have not been able to find a viable way to meet all their requirements, which continue to grow. One study, conducted by China Academy of Engineering (CAE), has projected that it will have a 10.5% energy gap in coming years and that SBSP systems may be one of the means to meet these needs [8].

Further, if energy needs are viewed from the perspective of national disasters, which in the past have led to the shutdown of energy-generating plants, then sunsats can be seen as a means to respond to power loss during emergencies. Further, this could also be seen from the perspective of national defense. It is important to have a source of supply if international energy sources are cut off or limited. In such a case it would seem vital to have sunsats as a potential source of supply.

Launch Systems and the Viability of Space-Based Power Systems

The viability of sunsats in the future heavily depends on the technical design of key components such as solar concentrators; the PV cells and their resilience to solar radiation; other similar technologies such as higher-efficiency quantum dot systems that can convert solar radiation into transmittable RF power; improved terrestrial rectenna design; and processes for finding large areas where these ground systems can be deployed. It is perhaps equally true that the development of improved and greatly more cost efficient launch systems are just as critical to completing the economic equation for designing, fabricating and deploying such systems in space.

In the chapter on launch systems, the significant progress that is being made in developing new and much more cost effective systems for deploying new space systems in orbit will be discussed. There has been significant progress in developing launch vehicles that can be used many dozens of times. These objectives are high on the priority list for new Space 2.0 firms such as SpaceX and Blue Origin. The current launch systems that are being designed for repeat missions, however, can only be used around 25 times. This is short of the studies that have suggested that as many as 100 reuses might be required to get the launch cost figures down to those projected to be needed – at least in some of the costing studies.

The key fact to note here is that the sunsats and the cost effective launch systems are integral to each other. They go together like a horse and carriage. You can't have one without the other.

Policy and Regulatory Concerns

The deployment of sunsats or SBSP systems will not only require the development of new technology and significant new capital investment but also give rise to a number of policy and regulatory concerns as well. The International Academy of Astronautics study of this subject, entitled "Assessment of Space Solar Power: Opportunities, Issues and Potential Pathways Forward," was conducted and reported on in 2011. This study suggested that the issues to be addressed and resolved included the following [9].

- What are the key wireless power transmission (WPT) beam health and safety considerations, and how should they be addressed in terms of health and safety standards and regulatory oversight?
- What spectrum allocation and orbital management procedures should be adopted to accommodate this service in terms of transmitting energy to Earth via radio frequency (RF) transmission or via other spectrum such as via laser transmission? (Currently the allocations that might be considered are 2.4 GHz and 5.8 GHz bands.) [10]
- What are the possible space debris impacts and related considerations in terms of debris mitigation and end of life removal from orbit?
- What might be the potential "weaponization" of the wireless power transmission systems associated with sunsats and what international agency might address these concerns?
- What is the best strategy and process for addressing the international coordination of these type systems in terms of their development and operations? [9]

The above listing is really only the start of the "what, why, where, how and who" of the issues to be considered. Here it is truly a case of "the Devil being in the details." There is a need for clarification within the International Telecommunications Union (ITU) as to the exact process that will be followed with regard to allocation of the exact frequencies that would be used for Regions 1, 2 and 3 or globally – most likely in the microwave band – or downlinking of the high-powered RF transmissions from a space-based system as well as for telemetry, tracking and control. There is likely to be a need to establish the power transmission limits that might apply, the process for assignment of exact orbital locations or characteristics for positioning of particular sunsats, and the process for frequency and interference coordination of such sunsats with communications satellites and other spacecraft.

There are literally dozens of issues, processes, standards and policies yet to be established. Will there be national licensing processes as well as registration of such sunsats with the ITU as well as with the United Nations, as required under the Registration Agreement? Further there are questions as to the liability agreement concerning how such systems are launched, owned, deployed and operated. This becomes particularly important if the sunsat is constructed in orbit and involves processing, manufacturing, assemblage, integration and quality assurance testing.

Conclusions

Virtually all of the space applications addressed up this point in the book involve well known practices and commercial applications where a good deal of precedent has been set as to how governmental licensing, due diligence and international collaborative processes should be followed. This is not the case so far as SBSP systems are concerned. Will such sunsats be operated as commercial projects by a single owner, by some sort of international partnership, or perhaps by some new type of international structure that follows such possible examples as an Intelsat, an Arianespace or some other model? There is truly a long list of health and safety standards still to be developed. Likewise there is a need for appropriate procedures with regard to registration, ownership, RF and orbital location coordination and more.

The bottom line is that there is still much more to develop and prove with regard to the viability and reliable performance of the technology. Notably missing are the needed launch capabilities as well as the technical and operational design of the SBSP systems themselves. There remains the need to validate the cost effectiveness of any currently deployable SBSP systems. This cost-effectiveness needs to be demonstrated not only in terms of competitiveness against traditional coal- and oil-powered electrical generation plants, but also against the ever more cost-efficient terrestrial renewable energy sources and energy storage systems that have increased significantly in the past decade, which are currently anticipated to increase in economic efficiencies by similar degrees in the coming decade as well.

References

1. Flournoy, D.M.: Solar Power Satellites*. Springer, New York (2012)
2. Landis, G.: Re-inventing the solar power satellite. TM-2004-212743, NASA Glenn Research Center. http://gltrs.grc.nasa.gov/cgi-bin/GLTRS/browse.pl72
3. Atkinson, N.: New company looks to produce space based solar power within a decade. Universe Today. https://www.universetoday.com/25754/new-company-looks-to-produce-space-based-solar-power-within-a-decade/ (2015)

4. Chapman, S.: Drop in solar and wind power costs to be driven by cheaper energy storage. Digital Energy. https://www.energydigital.com/renewable-energy/drop-solar-and-wind-power-costs-be-driven-cheaper-energy-storage (2018)
5. Optical Rectenna. https://en.wikipedia.org/wiki/Optical_rectenna (2018). Accessed 6 July 2018
6. Jakhu, R., Pelton, J.N.: Space-based solar power (Chapter 9). In: Jakhu, R., Pelton, J.N. (eds.) Global Space Governance: An International Study. Springer, Basel (2017)
7. Gopalaswami, P.A.: Sustaining India's economic growth on line journal of space communications. http://spacejournal.ohio.edu/issue/16/gopal.html (2010). Accessed 6 July 2018
8. Gao, J., et al.: Solar power satellites research in China on line journal of space communication. http://spacejournal.ohio.edu/issue16/ji.html (2010). Accessed 6 July 2018
9. International Academy of Astronautics (IAA): Assessment of Space Solar Power: Opportunities, Issues and Potential Pathways Forward. IAA, Paris (2011)
10. Op cit, Don M. Flournoy, p. 95

9 Space Weapons, the Threat of War in Space and Planetary Defense

Introduction

For fifty years the Outer Space Treaty of 1967 has provided a stable framework by spacefaring nations to carry out a wide range of activities in outer space. There have been some remarkable international cooperative ventures. We have seen the success of the Solyut-SpaceLab linkage, the International Space Station (ISS) and many other cooperative projects involving countries around the world. There is now an amazing array of cooperative organizations in the area of outer space. These include the International Telecommunications Satellite (INTELSAT) organization, with its service to almost 200 countries, the International Mobile Satellite (INMARSAT) organization, COSPAS-SARSAT, EUTELSAT, EUMETSAT, the International Academy of Astronautics (IAA), the International Space University (ISU) and many dozens of others.

Space has been a source of international cooperation, and the concept of space as a global commons has helped to promote peaceful cooperation in outer space.

Some, with a more cynical view, believe that the development of nuclear-armed weapons and missiles that can be targeted to specific locations anywhere on Earth via space navigational systems, such as the U. S.-operated NAVSTAR/GPS system, the Russian GLONASS system and so on have provided a modicum of world peace. This awesome framework of mutually assured destruction (MAD), as provided to the world by the visionary Hermann Kahn, has likely enforced the strong message that space should not be 'weaponized.' Certainly the fundamental concept of the Outer Space Treaty is that nuclear or other weapons of mass destruction were never to be deployed into outer space.

The basic principle that outer space, and Earth's orbital space in particular, would remain a nuclear-free zone and that space would be treated as a global commons for the benefit of humankind has brought a half-century of peaceful and cooperative uses of space. This has been a fundamental. It remains the agreed principle that all spacefaring

J. N. Pelton, *Space 2.0*, Astronautical Engineering,
https://doi.org/10.1007/978-3-030-15281-9_9

nations have willingly accepted in agreeing to the Outer Space Treaty and its four subsidiary international agreements. The Outer Space Treaty also provides the assurance that no one would seek to declare sovereignty over the Moon or any other celestial body and that we would keep weapons of mass destruction out of outer space.

Recent activities related to outer space, however, have raised new concerns about the possible militarization of outer space and Earth orbit in particular. Certainly the proposal by U. S. President Donald Trump to create a Space Force, even if never funded by the U. S. Congress, remains a topic of concern and discussion around the world. Some nations appear to have developed technology that might be considered as space weapons. These developments include high-powered laser or directed energy systems that could blind or disable vital satellites that provide key services such as communications, navigation, or Earth observation. Other developments have included spacecraft that can maneuver in orbit and are capable of approaching and disabling key satellites. At least three nations, China, Russia and the United States, have demonstrated capacities to launch anti-satellite weapons. The fact that North Korea is not a party to the Outer Space Treaty is of concern.

No one is sure as to what the exact purpose of a Space Force would actually be and whether this is different from the functions currently carried out by the U. S. Air Force. The key question is whether there is an implication that such a new entity would be responsible for deployment of either offensive or defensive weaponry in space.

In an entirely different vein, there are also concerns about non-manmade threats in space and about growing vulnerabilities that come with the addition of more and more people living in megacities who are dependent on modern infrastructure that might be at risk from planetary hazards.

There has been increasing research and new understanding into the nature of various cosmic hazards. We now know that potentially hazardous asteroids, solar flares, coronal mass ejections, and weaknesses in Earth's magnetosphere could put us at risk of enormous physical and economic danger if vital infrastructure is destroyed by cosmic hazards. Too often we forget we live on a large spaceship that is vulnerable in many ways. It is time to consider these vulnerabilities and to engage in what might be called planetary defense. But there is always a rub. Such defensive systems against cosmic hazards, if deployed in outer space, might also act as weapons systems in addition to defending our planet.

This chapter addresses the current status of space treaties and international agreements that define what various States have agreed to with regard to the use space for the common good and avoid conflict in space. It addresses capabilities that have developed or are under development that might be considered offensive or defensive space weapons systems. It examines what pending developments may be considered as contributing to the weaponization of space.

It thus addresses international concerns with regard to a possible future arms race in space.

Finally it discusses exactly what is being done to address both these types of concerns – weapons in space and planetary defense against cosmic hazards. It thus addresses the efforts of the U. N. Committee on the Peaceful Uses of Outer Space (COPUOS) and its Working Group on the Longer Term Sustainability of Outer Space Activities (LTSOSA), plus the U. N. Committee on Disarmament Affairs (CoDA) and of the U.N. General Assembly and Security Council.

The section on cosmic hazards and planetary defense addresses the significant related progress that has been made in this area, such as the recent creation of the International Asteroid Warning Network (IAWN), the Space Mission Planning Advisory Group (SMPAG) and further areas of progress, including efforts to address such issues as space weather, changes to Earth's magnetosphere, and efforts to address the problem of orbital space debris, which the latest Space Security Index Report identifies as the number one security issue.

Fifty Years of Progress Related to the Peaceful Uses of Outer Space

The world of space in the 1960s and 1970s was much different than it is today. The key terms of the Outer Space Treaty of 1967 were negotiated by the U. N. Committee on the Peaceful Uses of Outer Space (COPUOS), but in truth this was essentially a negotiation between the world's only spacefaring nations of the time – the United States and the Soviet Union. This was at the height of the Cold War, and neither nation wanted to see nuclear devices or weapons of mass destruction launched into space. The other members of COPUOS felt compelled to go along with the terms that the United States and the Soviet Union were able to accept. It is remarkable that in the space of only a few years the legal and regulatory framework of outer space activities were negotiated and agreed within the international community.

Not only was the 1967 Outer Space Treaty agreed but also the 1968 Astronaut Rescue Agreement, the 1972 Liability Convention, the 1974 Registration Convention and finally the 1979 Moon Agreement. The Outer Space Treaty has many key provisions, but perhaps none are more important than those of Articles I and IV.

Article I states: "The exploration and use of outer space, including the moon and other celestial bodies, shall be carried out for the benefit and in the interests of all countries, irrespective of their degree of economic or scientific development, and shall be the province of all mankind."

Article IV of the Outer Space Treaty contains the very important language that was intended to keep this region free of weapons. The first paragraph of Article IV explicitly states: "State Parties to the Treaty undertake not to place in orbit around the Earth any objects carrying nuclear weapons or any kinds of weapons of mass destruction, install such weapons on celestial bodies, or station such weapons in outer space in any other manner." [1]

Indeed Articles I, III, IV, IX and XI all express the basic premise of keeping the uses of outer space to peaceful purposes and to avoid the build-up of weapons in this key and strategic area [2].

Today the world of space has changed in significant ways. The number of spacefaring nations has risen to a dozen.

The number of members of the COPUOS has risen to over 80 countries plus over a dozen official observer organizations. The membership of the COPOUS has many members that hold widely different views on the best way forward in space cooperation that make achieving unanimous agreement extremely difficult – especially given its large and diverse membership. Most significant is the amount of space infrastructure that has been launched into space and is now operational for both civilian and defense-related purposes. Some of this space infrastructure is now so very vital that an attack on even the civilian space infrastructure such as space navigational, communications or remote-sensing satellites might be considered a declaration of war. The broad development of the ability of nations to deploy space infrastructure is provided in Table 9.1 [3].

Vital issues of concern now confront the COPUOS. Discussions within this forum sometimes involve questions as to the appropriate interpretation of the Outer Space Treaty and its four subsidiary international agreements. The definitive interpretation of U. N. treaties, of course, would ultimately lie with the International Court of Justice in The Hague, the Netherlands.

The areas of concern about how to maintain order and minimize conflict in outer space seem to continue to grow. Prime among such concerns are issues such as: (i) orbital space debris and active debris removal; (ii) deployment of large-scale satellite constellations and orbital congestion in LEO and GEO orbits; (iii) jamming and availability of RF spectrum; (iv) on-orbit servicing and potential reuse, salvaging or repositioning of defunct space objects; (v) space traffic management; (vi) cosmic hazards and planetary defense; (vii) the appropriate international policies and regulatory process that would apply to space resource acquisition; and especially

Table 9.1 Summary of space infrastructure deployed by spacefaring nations. (D = Defense and C = Civil) (Prepared by B. Jasani & R. Jakhu with some updates.)

Snapshot of the Civil- and Defense-Related Activities of Spacefaring Nations											
Country	Early Warning Sats	Comm Sats (D)	Comm Sats (C)	Nav Sats (D)	Nav Sats (C)	Met Sat (D)	Met Sat (C)	Obs. Sats (D)	Obs. Sats (C)	Indigenous Launcher	Space Weapons
China	X	X	X	X	X	X	X	X		X	X
No. Korea								X		X	
Europe		X	X	X	X	X	X		X	X	
France		X	X	X	X	X		X	X	X	
India			X		X	X		X	X	X	
Iran								X		X	
Israel			X					X	X	X	
Japan			X			X		X		X	
Russia	X	X	X	X	X	X	X	X	X	X	X
So. Korea			X						X	X	
USA	X	X	X	X	X	X	X	X	X	X	X

(viii) the creation of space forces, including the deployment of spacecraft that could be maneuvered in space and used as some sort of space-based weapon systems, and indeed possible deployment and use in space of military or defense systems.

Any type of space weapon system mentioned above could disrupt the current peaceful uses of outer space and international endeavors. A number of these issues are the concerns of the Working Group on the Long Term Sustainability of Outer Space Activities (LTSOSA) [4]

A half-century has passed since the unanimous international accords that were encompassed in the Outer Space Treaty were agreed to. The fast-moving and high tech world of space activities has seen remarkable changes. Tens of thousands of spacecraft have been launched. Orbital space debris has emerged as a significant problem. Private space ventures and commercial activities are now commonplace, even though this development was not anticipated in the Outer Space Treaty of 1967, nor the possible ownership of spacecraft recognized in the Liability Convention of 1972.

One would be amazed to think of a nation that has passed no new laws for fifty years – especially in a time of enormous technological progress. But this is the reality that applies to the international regulatory regime of outer space activities. In light of the seven important issues listed above and more, there is a need to find a way forward to continue to have a peaceful domain in outer space [2].

Too often there is talk of space wars and creation of space forces that might contend in Earth's orbital space, or beyond, or in the protozone, or sub space. This last is the area sometimes defined as from 20- to 160-kilometer altitude, where more and more new applications are being added each day. For the purposes of safety and of world peace some form of traffic management for Earth's orbital region and the protozone are becoming more and more urgent. The question is how can control or regulatory oversight for the protozone be accomplished? This is particularly a concern for applications that involve the extreme stratosphere and international uses such as hypersonic spaceplane flights across oceans. It will become an ever wider concern if there is no agreed mechanism for safe and peaceful regulation of this domain. Drifting balloons in the stratosphere with no controls, high altitude platform systems over international waters and international robotic air freight transport are just some of the contentious issues that may arise with regard to civil activities. It will be especially in the context of military or defense-related concerns and hypersonic spaceplane-based weapons systems that will most likely drive the most serious concerns. Currently there is considerable concern about what seems to some as a potential new arms race between the United States and Russia about hypersonic weapons and missile systems that would fly below outer space [5].

Soft Law Mechanisms to Supplement the Outer Space Treaty

A phrase that has arisen over the past decade or so is "transparency and confidence building measures" (TCBMs).

The basic idea is that nations, through a variety of mechanisms, would explain their various interpretations of international laws and regulations for space systems and how they would intend to act with regard to deployment of space systems in the future. The General Assembly actually established a Government Group of Experts (GGE) that was charged to come up with a recommended list of TCBM activities that States could follow that would contribute to disarmament in space and reduce the possibility of space hostilities. The recommendations of the GGE were presented to the U. N. General Assembly in 2013 [6].

There are some that are optimistic that the use of TCBMs can be of great value in the governance of space, particularly with regard to ameliorating the use of space for military or defense-related purposes. This attitude has been expressed as follows:

> *Transparency and confidence-building measures (TCBMs) are a set of tools designed to display, predict and discipline states' behavior with respect to maintaining the security of space. With intentional and unintentional threats to the peaceful use of space on the rise, there is a growing international consensus on the need for greater transparency in space-related activities as well as confidence-building measures to reduce the prospects of disruption to the ever-expanding role of space in our day-to-day lives* [7].

Others, however, feel that there are only 'admonitions' to good behavior associated with the TCBMs and that while these efforts and things such as Codes of Conduct are admirable they are really not truly effective unless there are enforcement provisions and some sort of watchdog capability attached.

There is in this regard the well-publicized European effort to establish an International Code of Conduct (ICOC) for outer space activities. This was considered by a number of nations and in discussions within the United Nations, but it has really not advanced in any significant new ways since 2015. Several of the reasons given for the lack of success by this effort have been the following: (i) the lack of any enforcement provisions or mechanism that would define when the ICOC had been violated; (ii) there was a lack of substance to the commitments other than an 'encouragement' for subscribing states to follow instruments already agreed to in the Outer Space Treaty and other documents seeking to limit space debris or armament of space; (iii) there were objections that this had not been done in a fully inclusive way and was essentially a European undertaking; and (iv) there are exceptions allowed on such bases as national defense that seemingly can always be argued by a State that wishes to use such a pretext for undertaking aggressive actions in space and then claiming defensive purposes [2].

The bottom line is that when this proposed European International Code of Conduct for outer space activities was taken up for discussion at the United Nations in 2015, the process ended up merely consisting of an exchange of views. During this U. N. discussion several participants indicated that the code drafting process had been largely based on European input. Although European delegations have indicated a determination to pursue this issue in 2016 through U. N. processes, no global consensus has yet developed, and the code remains only an encouragement to positive action [8].

There have also been proposals for new treaty provisions as well. China and Russia have put forward a draft Treaty on the Prevention of the Placement of Weapons in Outer Space. There is a companion proposal to a Treaty on Threat or Use of Force against Outer Space Objects (known as the PPWT). The most recent versions of these treaties were presented to the Conference on Disarmament in 2014. The discussion of these draft treaties has indicated flaws with verification processes of weapons that might have been launched, and indeed there are even problems with regard to a clear definition of what actually constitutes a space weapon.

Further there have been efforts to somewhat reverse the process of creating international standards related to disarming space. This has involved seeking adopting a U. N. Resolution that calls on spacefaring nations to unilaterally declare that they would not be the first to launch space weapons. This resolution, which is known as "Prevention of an Arms Race in Outer Space (PAROS)" is regularly passed in General Assembly sessions and notes that an arms race would be undesirable. Yet this PAROS resolution never includes any specifics on enforceability [9].

The Rising Threat of Militarization of Outer Space

It is often said by military officials who are concerned with such matters as national missile defense that space is congested, contested and competitive. As new weapons systems are developed and outer space is seen as a battleground where these systems might be used, the possibility of actual conflict in space seems to grow.

Each year there is a report entitled the Space Security Index. Its Executive Summary cites orbital space debris as the number one concern in this regard, but it also notes the serious concerns about the potential use of space weapons.

Here is a quote from its report for 2017:

> *......space security cannot be divorced from terrestrial security. In this context, it is important to point out that offensive and defensive space capabilities are not only related to systems that are physically in orbit, but include orbiting satellites, ground stations, and data and communications links. No hostile anti-satellite attacks have been carried out against an adversary; however, recent incidents testify to the availability and effectiveness of anti-ballistic missile systems to destroy satellites in outer space* [10].

Certainly China, Russia and the United States have tested anti-satellite missile systems, and a number of other countries, such as India, the United Kingdom, France and others, have undertaken research in this area (Fig. 9.1).

The development of various types of anti-satellite weapons systems using missiles with explosives, kinetic energy weapons, ground-based directed-energy systems and missile explosions to create electromagnetic pulses are numerous; no attempt will be made to catalog all these various systems, since such a listing would be out of date in a short period of time. It is sufficient to note that many different types of systems have been successfully developed by China, Russia and the United States, and other countries will undoubtedly follow.

Fig. 9.1 U. S. Vought ASM-135 anti-satellite missile launch on Sep. 13, 1985, that destroyed its P78-1 target. (Graphic courtesy of the Global Commons.)

Further it seems likely that if one country does proceed to create a space force a number of other countries will become motivated to follow course. The combination of such anti-satellite weapons systems and the continuing build-up of orbital debris represent a combined danger not only to world peace but to vital infrastructure in space on which the world depends operationally, economically, and in terms of national defense. Indeed the very future of so-called space security is tied to three very closely related issues. These are: (i) coping with and decreasing orbital space debris via mitigation techniques and regulations as well as active debris removal; (ii) establishing an international system for space traffic management (including the protozone as well as in Earth orbit) and (iii) curtailing current trends that seem headed toward the creation and perhaps use of various types of space-related weapons systems that include anti-satellite missile systems, spacecraft that could maneuver in space to attack another spacecraft, higher power directed energy beam systems, kinetic energy weapon systems and various types of nuclear weapons systems that might generate X-ray beams, electromagnetic pulses, or direct destruction of space systems.

In addition there is also mounting concerns about cyber-attacks on satellites, manned spacecraft, and spaceplanes and even missile systems. Today, according to a study by McAfee and the Center for Strategic and International Studies (CSIS), there are estimated losses incurred around the world amounting to over $600 billion. Such losses equal about 1% of the global economy, with over 25% of these losses incurred in the United States and perhaps 90% in countries of the Organization of Economic Cooperation and Development (OECD) [11].

The future development of such systems is clouded in uncertainty in many different ways. On one hand it is possible that some of the systems designed to cope with the reduction of orbital debris and to accomplish active debris removal could also be used as weapons systems. Likewise systems designed to cope with cosmic hazards, that include potentially hazardous asteroids or highly energetic space weather, could also be used as weapons systems.

It is likewise true that the development of new systems to cope with debris removal, on-orbit servicing or satellite repositioning could also be used as a form of space weapon. Ironically the development of new types of systems that might be used as space weapons could be considered for a very large new commercial and military space market. On the other hand the development of such capabilities might endanger the many commercial space applications now operating in orbit in areas such as telecommunications, networking, broadcasting, space navigation, remote sensing, Earth observation, and future markets such as on-orbit servicing, space-based solar energy, etc.

Sensible and effective new global space governance approaches are urgently needed to accentuate the positive opportunities and to block the most undesirable options that would involve warfare in space. Such a result could place billions of dollars, if not trillions of dollars, of vital space infrastructure at risk.

New Approaches to Space Traffic Management, Space Situational Awareness, and Space Safety

Currently the U. S. Joint Space Operations Center (JSPOC) keeps a close eye on all activities in space, including space debris orbital conditions and space launches. It keeps a special alert vigil with regard to missile launches and potential orbital collisions that might occur. It is ironic that JSPOC even alerts China as to possible collisions from the debris they created when they sent a missile up to knock out one of their old weather satellites in 2007. Such collisions create a problem not just for the owner of the affected spacecraft but for all spacefaring nations operating spacecraft in orbit. Everyone loses when a new collision happens in space.

Currently the cost of maintaining such a space watch is significant. The cost of the new S-band radar monitoring system that has been installed in the Pacific Ocean region was about $8 billion, and that does not include the annual operating costs. There is some planning for yet another S-band radar system that might be installed in Australia at a cost of many billions of dollars more. Other countries' space agencies are developing or considering developing their own radar or optical tracking systems as well.

There are, however, a number of key initiatives that are being undertaken by commercial space organizations at this time. A group of commercial space organizations with billions of assets in space formed an organization a decade ago known as the Space Data Association (SDA). This group's main mission is to be aware of possible conjunctions (or collisions) between operational satellites, as well as harmful frequency interference, and to assist with collision avoidance. It was founded by its four Executive Members: Intelsat, Eutelsat, SES, and Inmarsat. Its membership has grown steadily, and as of July 2018 had added the following members in addition to its founding Executive Members: Airbus, AMO-Spacecom, Arabsat, Avanti, Digital Globe, ExoAnalytics, GISTA of Thailand, Hispasat, NASA, NOAA, O3B, Omnispace, Optus, Orbcom, Planet, SSL, Spire, Telenor and Turksat. This organization is a unique mix of commercial remote sensing and communications satellite operators, governmental space and meteorological agencies and even commercial satellite fabricators. Its chief technology advisor is AGI, and the corporate advisor is MANSAT. SDA operates from the Isle of Man [12].

The Space Data Association reached an agreement with JSPOC to receive vital information with regard to possible conjunctions, and members of the Space Data Association receive continual information as to possible conjunctions between operational spacecraft and orbital debris of note.

There are many other efforts to track space debris and achieve improved space situational awareness. DARPA has reached agreement to deploy an optical telescope system in Australia to track orbital space debris [13]. Germany has developed a new radar system for debris tracking that involves a 34-meter radar dish at the Research Establishment for Applied Science (FGAN) at Wachtberg near Bonn, in Germany. There are also several commercial organizations that are deploying optical telescope and radar systems around the world to track debris and missile launches.

The following article by space situational awareness expert Brian Weeden in *Space News* summarizes the significant increase in commercial or commercially oriented tracking systems of orbital space debris: "The Space Data Association (SDA) provides participating operators with enhanced close-approach warning and radio-frequency interference resolution services. Several companies, including ExoAnalytics, Rincon, Lockheed Martin, and LeoLabs, are already selling SDA data from privately owned sensors. Other companies such as Analytical Graphics, Inc., Boeing, Schafer Corp., and Applied Defense Solutions are using commercial SSA [space situational awareness] data to create and sell services to governments and satellite operators." [14]

New commercial groups such as ExoAnalytics have approached the issue of creating a commercially viable global observation tool for monitoring space debris using a Space 2.0 mentality. Instead of billion-dollar massive installations they have created observational capabilities at 25 existing observatories and contracted with individuals at 200 sites to produce a massive real time database that they can integrate to form a global database [15].

Some, such as Weeden of the Secure World Foundation, have argued that others should take over monitoring the civil space situational awareness and have JSPOC concentrate on keeping defense assets secure. The key would be for JSPOC to follow the example that was set with the NAVSTAR-GPS network to allow commercial involvement with the operation and use of this network.

This type of shift of responsibility would involve such steps as: (i) working with the Space Data Association and key commercial providers of services to create 'a performance standard' for interfacing with the JSPOC network to obtain interoperability sufficient to support a civil space situational a service; create some form of "Interface Control Document" (ICD) that would allow machine-to-machine interoperability between domestic and international SSA capabilities that did not compromise the U. S. defense operational requirements to detect missile launches and threats to defense-related space assets [14].

Clearly there are opportunities to make space situational awareness and threat detection systems more effective and cost effective and allow commercial organizations to provide more of the services in these areas and make both defense and commercial systems better able to do their job and at lower cost.

Protecting Against Cosmic Hazards and Undertaking Planetary Defense

Many of the systems that are used to provide space situational awareness and to detect potential collisions between operational spacecraft and space debris are the very same systems that could be used to detect potentially hazardous asteroids and space threats. The problem is that there are many different types of cosmic hazards, and the scientific background and skills needed to address asteroid threats are quite different from those needed to detect intensive X-ray radiation flares from the Sun, coronal mass ejections and changes to Earth's protective shielding from the geomagnetosphere.

The U. S. Department of Defense, NASA, and other space agencies or defense agencies around the world have different units with different types of expertise working in geographically and functionally disparate groups to address different types of cosmic hazards. It is perhaps time for some fundamental questions to be asked about the importance, urgency and type of institutional support that is given to concerns about various types of cosmic hazards and how planetary defense issues are addressed, organized and funded on a national and global basis.

The first issue is why should there be more concern and priority devoted to cosmic hazards and planetary defense? The answer is that there are a great deal more people living in highly concentrated cities that are heavily dependent on modern infrastructure that are vulnerable to cosmic hazards and not easily or quickly replaced if the worst should happen. In truth, we are not prepared for catastrophic events that could wipe out the food and water supply, jobs and the livelihoods of billions of people if massive amounts of the vital infrastructure on which we now depend for transportation, supply chains, and economic survival were to go down. In 1700 there were 800 million people on Earth who were largely rural and sustained by farming, fishing, and enterprise not easily disrupted by cosmic hazards. Today we are headed toward a global population of over 9 billion by 2050 and which may be as much as 80% urban. These people are heavily dependent on global supply chains with jobs that are 80% services jobs. Their economic survival is based on electricity supply, water and fuel supply pipelines, industrial control systems and an urban infrastructure that is vulnerable to catastrophic loss from cosmic hazards.

For many years it was simply assumed that cosmic hazards were like the gods – beyond the reach and efforts of mere mortals to withstand their explosive fury. With the increasing sophistication of space systems, technology and launch systems it seems that there are indeed ways that humans could possibly design and create planetary defense systems that could defend spaceship Earth. Indeed with increased scientific and practical commercial space knowledge it might be possible to combine new scientific systems to learn more about the cosmos and practical systems with the commercial capability to produce space-based goods and services.

First of all it is important to have a better knowledge of things that go bump in the night and cause harm to humans. These can be cosmic hazards that are dangerous in a systemic way such as ultraviolet radiation and massive solar flares that create hazards to humans such as genetic mutation or skin cancer. They can be hazards that are dangerous to humans in the form of damage to infrastructure or create havoc in terms of a direct assault, such as a potentially hazardous asteroid hitting land or the oceans or bursting in the atmosphere or via a coronal mass ejection (CME) blasting Earth with trillions of high-speed ion particles. There are other types of dangers that include changes to Earth's magnetosphere that provides a natural protective shielding against CME blasts. This is a new type of hazard has only recently been identified through the ESA Swarm satellite constellation and the NASA MMS satellites.

Then there is the problem of a weakening of the ozone layer that creates various types of dangers to human health. Many would not identify the issue of climate change to be a form of cosmic hazard, but solar radiation and global warming are closely linked [16].

Human activities create space debris. This in turn serves to endanger our space-based infrastructure. Manmade changes to Earth's atmosphere and stratosphere serve to make some types of cosmic hazards even more dangerous.

In many ways it makes sense for space agencies and defense agencies to create a capability to look at and identify in a holistic and consolidated way all forms of cosmic hazards that are natural or induced by human activities. Only if we understand how all of these various cosmic dangers arise and potentially interact can we most efficiently defend humanity against these dangers. In terms of a strategic planning process, it seems that space agencies and defense agencies with space programs should put such objectives as their top priority. Scientific knowledge about space, the cosmos and how the cosmos evolved are important objectives, but survival of the human race should trump all other human activities in space and become job one.

There may be logical ways to create space-based solar shields to cope with coronal mass ejections and solar radiation, and compensate for changes to Earth's magnetosphere. It is significant to note that these types of solar shields could be used to build up a natural atmosphere on Mars, which does not have a magnetosphere. This type of consolidated and holistic thinking about planetary defense against cosmic hazards could also create new major industrial and commercial applications in space as well. There are design concepts for a planetary defense solar shield at Lagrange Point 1 that could be the focus of strategies to protect Earth from coronal mass ejections and mitigate the loss of shielding from the polar shifts of the magnetic north and south in coming years [17, 18].

The pursuit of such protective systems could lead to many new space-based industries as well. Some have suggested that industries related to the mining of asteroids could be closely linked to systems that use captured asteroids as a means to protect Earth from a hazardous asteroid. For example, an asteroid circling the Moon could be used as a shield to ward off such a threat.

Currently there are at least some tangible steps being taken in the United States to address near-Earth orbit threats. The approved five fundamental steps forward in a U. S. interagency report that addresses an action agenda with regard to NEO threats include the following:

- Enhance NEO detection, tracking, and characterization capabilities.
- Improve NEO modeling prediction and information integration.
- Develop technologies for NEO deflection and disruption missions.
- Increase international cooperation on NEO preparation.
- Establish NEO impact emergency procedures and action protocols [19].

Unfortunately efforts to address cosmic hazards are still piecemeal. A first step forward is thus to look at cosmic hazards as a whole. This would involve reorganizing space agencies and military space units to consider various types of cosmic hazards and planetary defense strategies as unified problems and then consolidate these efforts going forward. Such action might also create a pathway for increased international cooperation in such areas as combating orbital space debris, undertaking improved space situational awareness, and space traffic management activities for Earth orbit and the protozone.

Conclusions

Of all official governmental spending on space activities, expenditures on activities related to space weaponry and space situational awareness tops the list. U. S. military expenditures on space far outstrip the budget of NASA. The world of space today seems to stand at a crossroads. It is uncertain as to whether the future will be increasingly seen as "congested, contested and competitive" and with national space forces deploying new weapons systems in space, or will whether international tensions will subside and new opportunities will be found for space cooperation both by state governments and commercial ventures. There are a number of excellent opportunities for international cooperation – both governmental and commercial – to improve the effectiveness and cost efficiency in several related areas, such as space situational awareness, space traffic management, on-orbit servicing and active debris removal and planetary defense.

Of the various critical aspects of the future of space systems, it seems the perils and unrealized opportunities that exist in this area are truly the most important of all.

References

1. Treaties on Principles Governing the Activities of States in the Exploration and Use of Outer Space, including the Moon and Other Celestial Bodies, Entered into force as of October, 1967. http://www.unoosa.org/oosa/en/ourwork/spacelaw/treaties/introouterspacetreaty.html
2. Jakhu, R., Pelton, J.N.: Global governance of space security (Chapter 12). In: Jakhu, R., Pelton, J.N. (eds.) Global Space Governance: an International Study, pp. 267–300. Springer, Cham (2017)
3. Jasani, B., Jakhu, R.: Unsustainable uses of earth orbits. McGill Workshop on Bringing Earth Down to Earth, Montreal, Canada, July 4–5, 2013. www.mcgill.ca/iasl/events/2013-23f-mcgill. Accessed 10 July 2018. Note there have been some updates provided
4. Jakhu, R., Pelton, J.N.: Conclusions and the way forward (Chapter 23). In: Global Space Governance: an International Study, pp. 589–600. Springer, Cham (2017)
5. Button, K.: Hypersonic weapons race. Aerosp. Am. **56**, 20–26 (2018)
6. UN Doc. A/68/189
7. Robinson, J.: Transparency and confidence-building measures for space security. Space Policy. **37**(Part 3), 134–144 (2016). https://www.sciencedirect.com/science/journal/02659646
8. U.N. Document A/AC.105.1113, p. 8
9. UN Doc A/RES/69/32; U.N. Doc. A/RES/70/27; U.N. A/RES/71/32
10. Space Security Index, 14th Edition, October 2017. spacesecurityindex.org/wp-content/uploads/2017/10/SSI-Executive-Summary-2017-online.pdf
11. Palmer, D.: Cybercrime drains $600 billion a year from the global economy. February 21, 2018 says report. https://www.zdnet.com/article/cybercrime-drains-600-billion-a-year-from-the-global-economy-says-report/
12. Space Data Association. https://www.space-data.org/sda/about/membersandparticipants/. Accessed 12 July 2018
13. DARPA telescope headed to Australia to help track space debris. Nanowerk, 19 Nov 2012. https://www.nanowerk.com/news2/space/newsid=27529.php
14. Weeden, B.: Time for the U.S. military to let go of the civil space situational awareness mission. Space News, 20 Sept 2016. https://spacenews.com/time-for-the-u-s-military-to-let-go-of-the-civil-space-situational-awareness-mission/
15. Space situational awareness, Exo Analystics. https://exoanalytic.com/space-situational-awareness/
16. Pelton, J.N.: Space Debris and Other Threats from Outer Space. Springer, New York (2013)
17. Pelton, J.N.: Our changing world and the mounting solar risk of a calamitous solar storm. Room Space J. **2**(8), 62–66 (2016)
18. Pelton, J.N., Madry, S., Jakhu, R.: Defending Earth against cosmic hazards. Room Space J. **4**(6), 11–15 (2015)
19. NASA Press Release, National near earth object preparedness plan released, 20 June 2018. http://spaceref.com/asteroids/national-near-earth-object-preparedness-plan-released.html

Trends in Chemical Rocket Systems and New Approaches to Launching Satellites

10

Introduction

The world of satellite applications has been turned upside down for several reasons in the past decade. The advent of small satellites and large-scale constellations, high throughput satellites, new types of Earth stations employing metamaterial and electronic beam forming, and new launcher design and operation – including reusable vehicles – have all served to reinvent the world of space. The common factors these innovations share are innovative thought and entrepreneurial initiative. The significant changes that have come to the launch industry, on which the broader space industry depends, is truly a key part of the reinvention of the space industry. This chapter examines the most important of these changes and also sets the stage for an exploration of the further changes yet to come, which will be explored in chapter eleven.

In the world of spacecraft design, the shrinking size of high-speed digital processors and miniaturized sensors have allowed the satellites of today to shrink in size while increasing performance. Spacecraft have become smarter, more capable and more cost effective. This shrinkage in size with increased performance has been a good deal of what the reinvention of the satellite industry has been about. When computers get smaller they get faster and more capable. Satellite designers have found more economical ways to perform the same trick.

The reinvention of launcher systems has been the other half of the equation. We are now seeing better launchers that are better adapted to the needs of Space 2.0 spacecraft systems. We are especially seeing the development of new capabilities such as reusable rocket systems, more efficient manufacturing techniques and clever new ways to launch rocket systems that do not require the operation of expensive rocket launch sites such as were an essential part of the rocket industries' infrastructure. Paul Allen and Vulcan, Inc., have now developed the massive Stratolaunch aircraft that can fly out of an airport and carry a massive rocket up to high altitude for quite cost-effective launches. The world is not like it used to be.

J. N. Pelton, *Space 2.0*, Astronautical Engineering,
https://doi.org/10.1007/978-3-030-15281-9_10

The Evolutionary Design of Launch Vehicles and How They Work

The idea of rocket propulsion is actually not at all new. Just recall the story of Archytas of Tarentum, 400 to 350 B. C. He was particularly noted for his steam-powered pigeon that flew around in a tethered circle on a rope inside of his home. The idea of heating some sort of fuel that could be expelled through a jet to create a propulsive force was thus documented to have taken place 2400 years ago [1].

The Chinese creation of gunpowder led to the pyro-techno rockets of the 14th century. There is perhaps the apocryphal story of the Chinese nobleman who aspired to be the world's first astronaut. Supposedly he sat astride a giant array of 144 rockets for a journey into space, but his reward for his efforts, unfortunately, was to be instantly immolated. If true it indicates that the aspiration to fly into space on a rocket-powered vehicle is a concept that has long been with us.

The first rockets thus were solid-fueled. These were built by the Chinese some 900 years ago and were reportedly first used in warfare in 1232 in the Chinese war with the Mongols. They were then, as they are today, essentially rocket-shaped bombs [2].

These rockets were first fueled by gunpowder and then by other explosive compounds. The compounds have become more sophisticated, but the principles remain the same. The designs have involved placing various explosive solid-rocket fuels within a chamber that allowed continuous ignition and expulsion from a rocket cone. There was no opportunity for stopping the ignition once started, since this was, in effect, a continuously firing bomb. Solid-fueled rockets as they exist today are well-suited for missile weapon systems. This is because they can be instantly fired at any time without the loading of the rocket with fuel.

The liquid-fueled rocket came later. The idea of liquid fuel being pumped into a combustion chamber in a controlled manner was perhaps first realistically envisioned by the Russian Konstantin Tsiolkovsky in the late 19th and early 20th centuries [3].

The first actual liquid-fueled rocket system was developed on a rudimentary level by the American rocket scientist Robert Goddard in the 1920s and 1930s. For his efforts, *The New York Times* ridiculed him and called him "The Moon Man." [4] When Neil Armstrong and Buzz Aldrin walked on the Moon nearly a half century later, *The New York Times* offered Goddard a formal apology for making fun of his prediction that liquid-fueled rockets would allow people to walk on the Moon someday.

There are several efficiencies that can be derived from liquid-fueled rockets. These advantages include the fact that they can be throttled, turned on and off and produce less pollution, especially regarding particulates. The most explosive liquid fuel combinations, such as liquid hydrogen that is oxidized by liquid oxygen, generates greater thrust as well. The basic aspects of the operation of a liquid-fueled rocket plus key formulas of physics concerning their propulsion systems are shown in Fig. 10.1.

These types of liquid-fueled rocket systems do require complicated fueling, pumping and valve systems. Those that use liquid oxygen and liquid hydrogen, for instance, also require special and

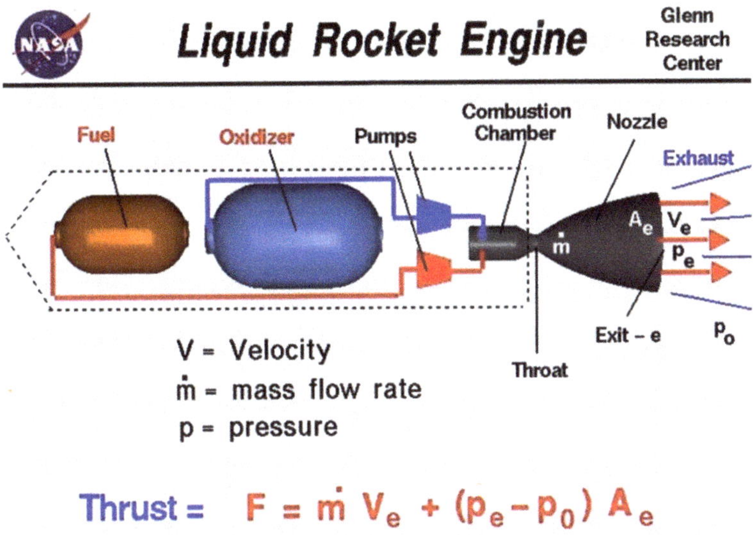

Fig. 10.1 The basic concept of a liquid-fueled rocket. (Graphic courtesy of NASA.)

expensive refrigerant systems. These types of rockets are thus best suited to scientific or commercial missions, where there is no need for an instant response, such as to fire off a weapon.

Today there are also so-called hybrid systems. These types of rocket systems might have a solid fuel but are oxidized by a gas or liquid that can be throttled. This control feature makes this type of propulsion system suited for such applications as spaceplanes, which have human crew aboard. The first such workable system was developed by the Benson Space Development Company and included the seemingly unconventional fuel of neoprene rubber that was oxidized in a controlled fashion with laughing gas (i.e., nitrous oxide). The supply of nitrous oxide, which could be cut off, served as the throttle oxidizer.

Other hybrid rocket propulsion systems are now being used in the SpaceShipTwo spaceplane to achieve a higher efficiency level of propulsion. The ability to throttle off the oxidizer is a key safety feature, but it is still true that all solid-fuel rocket systems produce high levels of pollution and spew particulates into the stratosphere.

It is especially important to note the particular effects of pollutants in the stratosphere. In these high altitudes, where the atmosphere is perhaps 100 times less dense than at sea level, means that the adverse environmental effects are much, much greater when particulates are released into the stratosphere.

Ion Propulsion and Electrical Space Vehicles

The most recent area of development for rocket propulsion involves electric propulsion. This approach uses electronic guns to accelerate ions.

Electric propulsion is now used to achieve more cost efficient and longer-lived station-keeping for spacecraft, particularly for those operating in GEO orbit. It typically uses ionized xenon fuel. This type of approach has become widely used for spacecraft system control, but the motors do not create enough thrust for launching satellites into orbit. In short, this type of system creates thrust and accelerative forces for much longer periods of time than a chemically fueled rocket. They can last hundreds or thousands of times longer, but there is not enough thrust – or concentrated surge – to overcome the pull of gravity at sea level.

There is some thought that ion propulsion might be able to lift a small satellite – like a cubesat – to orbit from a dark sky station positioned many kilometers high in the stratosphere. Ion propulsion can be more efficient and less polluting. The first type of electrical propulsion system was developed for lower thrust station-keeping and Vernier-jet orientation systems for spacecraft (See Fig. 10.2, which is a functional diagram of a gridded ion thruster that uses xenon fuel).

The thrusting power of ion propulsion, as just explained, is much too low to support launching operations. However, these systems are very fuel efficient and could allow a spacecraft to be maintained in a geosynchronous orbit with over a particular location for many years and do so with a reasonably small amount of xenon gas as the fuel. An electronic gun is used to electrically ionize the xenon gas to create low thrust levels for needed station-keeping operations. Despite the thrust levels being low, the overall net performance

Fig. 10.2 Artist conception of the xenon-fueled gridded electronic ion thruster. (Graphic courtesy of NASA.)

over time is much higher. In testing of the NASA NEXT electric ion propulsion system, which ionizes xenon gas inside of its chamber and then emits them at very high speed, the thrust is impressive.

The net total measured effective propulsion is 12 times higher than a chemical rocket propulsion system when it is measured over time and a comparison is made of fuel consumed. It is thought that in time such systems, with perhaps a nuclear power supply for the electronic gun, could be used to support interplanetary missions. The thrust levels are low but constant. Thus great speeds could be built up over the span of months for such missions traveling to outer reaches of the Solar System [5].

New Approaches to Design and Manufacture of Launch Systems

The basic approaches for designing rockets and missile systems are now well known, but there is still room for improvement in how these systems are designed and manufactured. These improvements do not have to involve a new type of propulsion. The key is finding better ways to design, manufacture, undertake quality testing and even ways to build reusable rocket systems. In the manufacturing there is the potential to use 3D printing or additive manufacturing. There are some that envision using additive manufacturing to create rocket engines and other key components for rocket launchers. This approach might not only reduce the cost of producing the motors and the components but also could help streamlining and simplifying the quality assurance assets of the manufacturing process.

There is an even more fundamental shift now in process, and that is the changeover from expendable launch vehicles (ELVs), which can be used only one time, to launchers that might be reused from twenty to even thirty times. These design concepts require that first stage vehicles return to predetermined locations. This technology has been demonstrated at least in early stages by both Elon Musk's SpaceX and Jeff Bezos's Blue Origin. If these two reusable launch programs are successful, then other programs will undoubtedly follow.

Yet another aspect of the SpaceX launch program's effort to achieve greater cost efficiency is the drive to vertical integration. Thus SpaceX is seeking to build its launchers motors, fuel tanks, pumping and refrigerating systems and other components so that it can control supply chains and optimize its production to both ensure quality and to control costs.

In design concept anyway, rocket systems that might be used over and over again could serve to reduce the cost of launches by a very meaningful degree. This could be the most significant change in reducing costs sufficiently to make large-scale solar power satellites viable and make the assemblage in space of other large-scale structures for research or to create a 'Sun shield' against large-scale coronal mass ejections during the period when Earth's magnetic poles are shifting from north to south and south to north. Such new launch economies could also facilitate the creation of permanent colonies on the Moon or Mars.

Reusable launch vehicles might also allow for the creation of larger scale structures and habitats in space that might be used for everything from space tourism to various forms of planetary defense and even create structures in space to cope with climate change. In short, it could be a whole new ballgame.

However, we must temper our sense of enthusiasm to consider the environmental effects of a major increase in chemically fueled launches to orbit and the potential for pollution of the stratosphere. In short, reusable launch vehicles might be best seen as a transitional step from chemically fueled launch vehicles to new systems that might use tethers or even space elevators to lift mass to orbit perhaps four or five decades from now. It is important not to view the future through a rear view mirror. It is probably good to think that there are better ways to put people and satellites into space than putting them on top of a controlled bomb.

Launch Systems for Cubesats and Small Satellites

There are systems such as the largest Arianne vehicle that will be able to launch as many as 35 of the OneWeb small satellites (250 kg each) all at once, where economies of scale are clearly at work. Yet for emergency restoration purposes, the option of an efficient small scale launch becomes apparent. The Virgin Galactic Corporation is not only developing SpaceShipTwo for suborbital flights but another carrier vehicle that would lift Launcher One to 50,000 ft (about 14 km) for launch. The current design for Launcher One would allow it to launch two OneWeb small sats at a time. This would give OneWeb the opportunity to be able to respond quickly to satellite outages (See Fig. 10.3). The use of carrier vehicles as

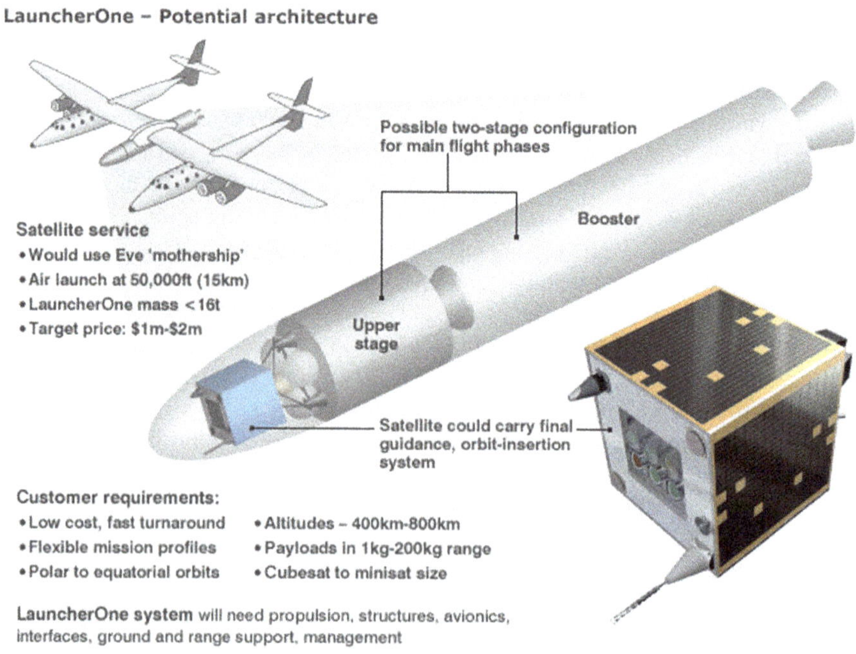

Fig. 10.3 Launcher One provides new small sat capabilities. (Graphic courtesy of Virgin Galactic.)

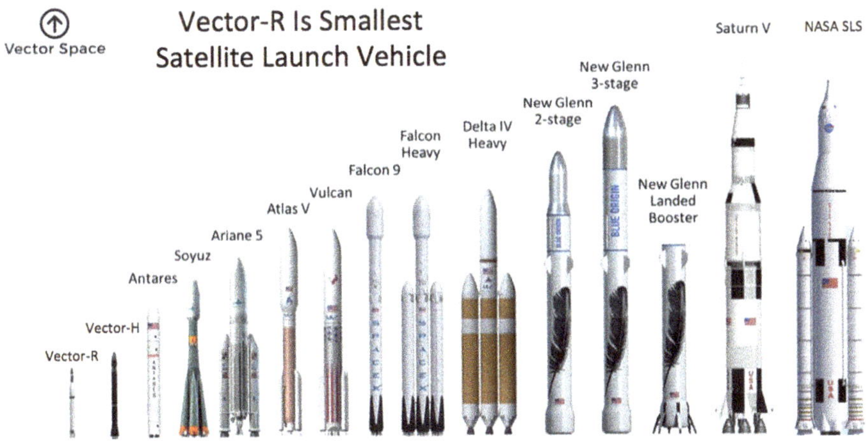

Fig. 10.4 The range in the size of launch vehicles keeps expanding. (Graphic courtesy of Vector Space.)

a part of the first stage of a reusable launch system will be discussed in more detail below.

There are now a number of small-scale launchers under development that will be able to launch cubesats and larger small satellites with good efficiency. Some of these options include the Vector R and Vector H Launchers. Figure 10.4 provides perspective on the huge range in size and lift capability that is now available to accommodate spacecraft that today might range in size from 100 g–10,000 kg. It should be noted that new configurations that allow launchers to accommodate many small satellites on a single launch adds great flexibility. The Indian Polar Satellite Launch Vehicle managed to launch two medium-sized spacecraft and 104 cube satellites in a single launch. This represented a new record for the number of satellites launched on a single launcher.

Carrier Vehicles and Spaceplanes as First-Stage Ascent Systems

The effort to develop spaceplanes as a way to provide for reliable suborbital flights for so-called space tourists could be said to have given rise to the idea of reusable launch vehicles. As companies developed plans to build spaceplanes such as SpaceShip2 by Virgin Galactic or the now defunct S-3 spaceplane, their business objectives grew to include the launch of small satellites to low Earth orbit. In some scenarios, it was also planned to have reusable spaceplanes at the first stage of the launch.

Thus, as noted above, Virgin Galactic is now offering Launcher One services to launch small satellites. The Launcher One will fly on a carrier vehicle called Eve. The S-3 spaceplane, however, had been envisioned to fly on a modified jumbo jet that would have served as the first stage of a small satellite launch system. The lift system to get quite small payloads to orbit would then be the final stage that would launch from the S-3 spaceplane. This ambitious venture has now gone bankrupt.

At the other end of the scale from the Launcher One and the S-3 small payload to orbit system is the plan of Vulcan Industries. Vulcan is now the sole developer of the Stratolaunch system, which is designed to serve as a reusable carrier vehicle for much larger rockets. The Stratolaunch carrier vehicle is powered by six 747 engines and is being designed to take much larger rocket launchers to high altitude for launch. Again, the thought process is that if one can reuse the first part of a launch system over and over again it becomes much more cost effective.

The traditional approach to launching rockets into orbit has logically involved the creation of launch sites. The creation of such launch sites have resulted in the construction of launch gantry towers, systems to load fuel into rockets and control rooms where key components on the launcher can be monitored by launch operation engineers and scientists. These launch sites take up a fair amount of room in order to accommodate all these functions and to isolate them from populated areas for safety and security reasons. The Kennedy Space Center in Florida in the United States, the Space Center in French Guiana, or the Baikonur Cosmodrome launch facility first built by the Soviet Union, are indicative of how expensive these launch sites were to design and build up into the giant complexes they are today.

As innovations have been made over time to create more reliable launch systems and innovators have sought to find more cost-effective ways to operate them, part of the focus has been on how to launch at points of maximum efficiency (such as along the equator for geosynchronous orbit launches) and how to avoid the high cost of launch facilities. One approach that has been taken is that of a sea launch. This approach was to have a mobile and seaworthy launch platform in the ocean (operating out of southern California) that could take a rocket and launch it from the high seas near the equator. A wide range of other options have been explored. One idea is to lift the rocket up with balloon systems that would launch from the upper atmosphere. Another concept is to have a towing system to haul rockets or spaceplanes to a high altitude for launch. One idea that has been in practice for some time is to have a carrier vehicle that would transport a rocket to a certain altitude and release it for mid-air ignition. This approach was used by Orbital Space Systems for the Pegasus and Taurus vehicles. As already noted, this is also fairly similar to the approach that is being used by Virgin Galactic with Launcher One and with Stratolaunch for a larger launcher, and was used for the now defunct S-3 spaceplane system.

The other advantage that comes with such an approach is that it is simply a more efficient way to overcome the gravitational pull of Earth's gravity well. This decreases as one achieves higher altitude. A rocket launch that occurs at perhaps 14 or 15 km high requires somewhat less fuel to reach orbit. Also a launch to GEO orbit from the Kennedy Space Center has a 14% disadvantage as compared to a launch from the Kourou Space Center in French Guiana because of the relative

Fig. 10.5 The rollout of the Stratolaunch by Vulcan Industries – the world's largest plane. (Graphic courtesy of Vulcan Industries.)

rotational speed of Earth at these respective latitudes.

It was the intention of Microsoft founder Paul Allen, who is the owner of Vulcan Industries, that his largest plane, the Stratolaunch, be a boon to rocket launcher companies (See Fig. 10.5). His company hopes to offer rocket companies the ability to launch large rockets at low cost and high efficiency because: (i) they would not have to pay the high cost of creating, maintaining and licensing a ground-based facility and scheduling a launch at such a facility; (ii) there would be fuel and safety advantages of launching from high altitudes in the stratosphere; and (iii) the launch could take place at exactly the longitude and latitude location desired, such as along the equator for GEO orbit launches. These same cumulative advantages also accrue to similar carrier launch systems, such as the Launcher One.

Launch Sites Around the World

There will remain a significant number of launch sites that are from ground locations. Most spacefaring nations maintain one or more launch complexes in their own country. There are several launch sites, such as those of the European Space Agency in French Guiana and the ISRO launch site in India, that have been established at locations that are attractive because of proximity to the equator. Others have been chosen due to their suitability for polar-orbit launches or because of their isolated or shore locations. This is because it provides a useful safety measure in case of a rocket malfunction or accident involving rocket fuel combustion. The following list represents significant launch sites currently operating around the world.

Australia

Site Name: Spaceport Australia
Location: Woomera, Australia (Latitude 31.1°S Longitude 136.6°E)
Launch Vehicles Supported: None currently active

Site Name: Asia Pacific Space Center (proposed)
Location: Christmas Island, Australia (Latitude 10.4°S Longitude 105.7°E)
Launch Vehicles Supported: Aurora (proposed)

Brazil

Site Name: Alcantara Launch Center
Location: Alcantara, Brazil (Latitude 2.3°S Longitude 44.4°W)
Launch Vehicles Supported: VLS-1 (proposed)

China

Site Name: Jiuquan Satellite Launch Center (JSLC)
Location: Gobi desert, Inner Mongolia (Latitude 40.6°N Longitude 99.9°E)
Launch Vehicles Supported: Long March 2C/2D/2F & Long March 4B/4C

Site Name: Xichang Satellite Launch Center (XSLC)
Location: Xichang City, China (Latitude 28.3°N Longitude 102.0°E)
Launch Vehicles Supported: Long March 2C & Long March 3A/3B/3BE/3C

Site Name: Taiyuan Satellite Launch Center (TSLC)
Location: Shanxi Province, China (Latitude 37.5°N Longitude 112.6°E)
Launch Vehicles Supported: Long March 2C/2D and Long March 4B/4C

Site Name: Wenchang Satellite Launch Center (WSLC)
Location: Hainan Island, China (Latitude 19.7°N Longitude 111.0°E)
Launch Vehicles Supported: Long March 5 (proposed)

Europe

Site Name: Guiana Space Center (Centre Spatial Guyanais)
Location: Kourou, French Guiana (Latitude 5.2°N Longitude 52.8°W)
Launch Vehicles Supported: Ariane 5, Soyuz, Vega, and Ariane 6 (Proposed)

India

Site Name: Satish Dhawan Space Center (SHAR)
Location: Sriharikota Island, India (Latitude 13.9°N Longitude 80.4°E)
Launch Vehicles Supported: Polar Satellite Launch Vehicle (PSLV) and Geosynchronous Satellite Launch Vehicle (GSLV)

Iran

Site Name: Iranian Space Agency Emamshahr Space Center, where suborbital LV have been launched. Qom, below, is the other launch site
Location: Emamshahr Space Center located at 36°25′0″N 55°01′0″E
Launch Vehicles Supported: Shahab 3

Site Name: Qom Space Center
Location: Qom, Iran, located at 34°39′0″N 50°54′0″E
Launch Vehicles Supported: Shahab 3

Israel

Site Name: Palmachim Air Force Base
Location: Negev Desert, Israel (Latitude 31.5°N Longitude 34.5°E)
Launch Vehicles Supported: Shavit

Japan

Site Name: Tanegashima Space Center (TNSC)
Location: Tanegashima, Japan (Latitude 30.4°N Longitude 131.0°E)

Launch Vehicles Supported: H-IIA and H-IIB

South Korea
Site Name: Naro Space Center
Location: Goheung County, South Jeolla (Latitude 34.4°N Longitude 127.5°E)
Launch Vehicles Supported: Naro-1

Russia
Site Name: Baikonur Cosmodrome
Location: Tyuratam, Kazakhstan (Latitude 45.6°N Longitude 63.4°E)
Launch Vehicles Supported: Proton, Strela, Dnepr, Zenit, Rockot, and Cyclone 2 (In addition the Soyuz and Vega are launched from the Guiana Space Center as noted above.)

Site Name: Plesetsk Cosmodrome
Location: Arkhangelsk Oblast, Russia (Latitude 62.8°N Longitude 40.1°E)
Launch Vehicles Supported: Kosmos 3M, Rockot, Soyuz, Start-1, Angara

Site Name: Svobodny Cosmodrome
Location: Amur Oblast, Russia (Latitude 51.4°N Longitude 128.3°E)
Launch Vehicles Supported: Start-1 and Rockot

United States
Site Name: Cape Canaveral Air Force Station (CCAFS)
Location: Cape Canaveral, Florida (Latitude 28.3°N Longitude 80.3°W)
Launch Vehicles Supported: Falcon 9, Atlas V, Delta IV
Site Name: Kennedy Space Center (KSC)
Location: Merrit Island, Florida (Latitude 28.5°N Longitude 81.5°W)
Launch Vehicles Supported: Space Shuttle (retired), Space Launch System Constellation

Site Name: The Mojave Spaceport
Location: California, USA (Latitude 35.0°N Longitude 118.2°W)
Launch Vehicles Supported: Various horizontal takeoff spaceplanes

Site Name: Spaceport America (formerly known as the Southwest Regional Spaceport)
Location: Las Cruces, New Mexico (Latitude 32°N Longitude 107°W)
Launch Vehicles Supported: White Knight Two Carrier Plane and SpaceShip Two

Site Name: Vandenberg Air Force Base (VAFB)
Location: Lompoc, California (Latitude 34.4° N Longitude 120.35° W)
Launch Vehicles Supported: Delta II, Delta IV, Atlas V, Minotaur I, Minotaur IV, Taurus, Pegasus, Falcon 1

Site Name: Wallops Flight Facility (WFF) and adjacent Mid Atlantic Spaceport of the State of Virginia and the State of Maryland
Location: Wallops Island, Virginia (Latitude 37.8°N Longitude 75.5°W)
Launch Vehicles Supported: Pegasus, Minotaur, and Antares launch vehicles of Orbital ATK

Site Name: West Texas Test Facility for Blue Origin
Location: West Texas near the New Mexico State Line
Launch Vehicles Supported: Latest is the New Shepard. New Glenn Operations to be moved to Florida along with a new Blue Origin plant (See Fig. 10.6).

Note: Several dozen other U. S.-based spaceports (essentially all for horizontal takeoff and landing operations) have been licensed or have licensing pending, but none is currently operational, with

Fig. 10.6 New Shepard reusable vehicle preparing for launch at west Texas site. (Photo courtesy of Blue Origin.)

actual spaceplane operations. Several other spaceports in locations such as the United Kingdom, Singapore, Malaysia, Sweden and Italy, among others, are now anticipated. Recently a framework agreement was signed by Virgin Galactic and Italian companies Altec and Sitael. Under this agreement the two companies will continue planning for potential flights of Virgin Galactic's SpaceShipTwo from the Taranto-Grottaglie airport in the southern part of Italy [6].

Many of the above-mentioned launch sites are equipped with launch gantries and also contain special refrigerant facilities for liquid oxygen and liquid hydrogen and other specialized facilities, test firing systems for engines and system integration equipment. Many of the launch facilities located around the world are also designed to allow missiles or rockets to be launched from these locations as well. There are a growing number of spaceports that have been or are being built or licensed, but few are operational in that the spaceplanes that would use them are still at the research, development or early testing stage. Virgin Galactic and Blue Origin are among the first of these spaceplane and small satellite launching enterprises, but others will undoubtedly follow in the near future.

There are many missile silos for weapons systems. Further there are also missile systems on ships and submarines as well as missiles that can be launched off of mobile platforms such as specially equipped trucks and trains. These types of missile systems, however, are virtually all solid rocket systems that do not involve loading of liquid fuels.

There are also many spaceports that are quite diverse in their size, sophistication and safety specifications.

The capabilities vary from airports that have suitable runways that can basically only accommodate horizontal take-off and landing and yet get authorized by their governments to be called "a spaceport." At one time there was the thought that spaceports had to be limited to locations with takeoff and landing over an ocean, for safety purposes, but this restriction no longer applies in most countries. The location is, however, frequently limited to either an ocean adjacent location or isolated areas.

Essential Ground Support Systems for Launch Operations

It might be easy to assume that launch operations are all about the launchers and their design and operation. It is important to note that these systems do not operate without systems on the ground for tracking, telemetry and command of the launch vehicles as they are launched into space, achieve specific orbits and deploy spacecraft.

There are a number of command and control centers for satellite networks that are also used to support launch operations. Specific commands are needed to launch operations, and these can come at different locations and altitudes all around the world. Just as many satellites are moving toward autonomous operations, launch vehicle systems can be preprogrammed to carry out specific functions such as the cutoff of engines or re-ignition of engines to achieve specific orbital conditions and altitudes.

It is important to maintain the ability to send critical commands to cope with such happenstances as a misfiring of a rocket motor. This might even include the need to execute a command to destroy a launcher if the rocket flies off course and threatens a community. Manned missions, in particular, require the ability to stay constantly in radio contact with a rocket at all points in its operation.

Stratospheric High Altitude Platforms: The Launch of Pseudo-Sats

The advantage of satellite systems is their very high altitude, which provides them with a very broad field of view. The higher the altitude, the greater area that can be surveyed by remote-sensing satellites or the greater the number of communications stations that can be connected. It is more difficult to launch a satellite into geosynchronous orbit because it requires greater thrust, but then it requires only three satellites at the high altitude to essentially cover the entire Earth. In low Earth orbit, of say 800 km, it takes perhaps 60 satellites to cover the entire Earth because of the lesser coverage, but each of these launches require less fuel and thrust to achieve orbit. Until recently the options for remote sensing or communications were towers, aircraft, aerostats or the deployment of additional satellites to achieve much broader coverage.

More recently there has been attention given to the concept of high-altitude platform systems (HAPS) and unattended aeronautical vehicles (UAVs) as means of achieving broad coverage that is greater than towers but less than that of satellites. Frequencies have been allocated for such service, and an increasing number of studies are being made as to how such systems would be managed in terms of safety and flight traffic management. HAPS are being designed to operate in the stratosphere for longer term operation. Thales Alenia, for instance, has recently signed an agreement with the Southwest Research Institute to develop the Stratobus HAPS system [7].

These platforms are sufficiently elevated in altitude that they can provide coverage for island countries such as Jamaica, Fiji, or even Iceland. Because of their lower altitude, the path loss is minimal. Thus such HAPS systems can have high digital throughput capabilities.

These systems are varied in their design in terms of their maneuverability and stability. Some that involve lighter than air dirigibles can stay aloft for sustained periods of time. Some are conceived as automated jet aircraft that have to be periodically refueled, while others are designed as solar-powered craft with electric motors that can stay up for sustained periods of time. Yet others are simply stratospheric balloon systems that depend on global wind current conditions.

There are today concerns about not only space traffic management and how to control spacecraft and debris against collisions, but also about safe operation of HAPS and other things that might be designed to operate high in the upper stratosphere in future years. There could also be environmental concerns about systems that are designed to fly in the stratosphere that would require the use of expendable fuel systems such as those that would use jet engines for propulsion.

Conclusions

There have been significant improvements in the design and operation of various types of launch systems in the past decade. Innovations that have allowed launch vehicles to become more reliable and progress to create reusable launch systems have shown a pathway to much more reliable and cost-effective launchers. New techniques such as 3D printing and additive manufacturing have also contributed to lower costs of the production of launch systems.

For environmental reasons, rocket launches for commercial operations are heavily geared toward the use of liquid-fueled vehicles. For reasons of safety, launch operations are designed to clear all aircraft flights in proximity to rocket launches. There is the thought that with the advent of space traffic management it might be possible to integrate air traffic management and control systems and space traffic management systems in the future.

Key aspects to be focused on for the future are new manufacturing and design innovations to reduce the cost of launch operations while also making launches safer. The advent of more launches of large-scale constellations into orbit raises new concerns with regard to orbital space debris and also with regard to pollution of the stratosphere.

References

1. Archytas of Tarentum, Encyclopedia Britannica. https://www.britannica.com/biography/Archytas-of-Tarentum
2. A brief history of rockers, NASA Glenn Research Center. https://www.grc.nasa.gov/www/k-12/TRC/Rockets/history_of_rockets.html. Accessed 18 July 2018
3. Redd, N.T.: Konstantin Tsiolkovsky: "Russian Father of Rocketry". Space.com (2013). https://www.space.com/19994-konstantin_tsiolkovsky.html

4. Robert Goddard: The moon man. Legacy.com. http://www.legacy.com/news/explore-history/article/robert-goddard-the-moon-man. Accessed 18 July 2018
5. NASA NEXT ion drive breaks world record. http://www.extremetech.com/extreme/144296-NASA-next-ion-drive-breaks-world-record-will-eventually-power-interplanetary-missions. Accessed 15 July 2018
6. Foust, J.: Virgin space companies sign new agreements with Italy. Space News (2018). https://spacenews.com/virgin-space-companies-sign-new-agreements-with-italy/
7. Thales Alenia Space and SWRI sign MOU to cooperate on stratobus development (2018). https://www.thalesgroup.com/en/worldwide/space/press-release/thales-alenia-space-and-swri-sign-mou-cooperate-stratobus-development

11 The Longer Term Future of Launch and Propulsion Systems

Introduction

In the previous chapter the remarkable progress that is being made to increase the reliability, performance and cost effectiveness of launcher vehicles was presented in some detail. This particularly stressed the progress that is being achieved with reusable launch systems. SpaceX Falcon launchers, Blue Origin New Shepherd and planned New Glenn and Sierra Nevada's Dreamchaser vehicle that would provide resupply services to the International Space Station are just some of the new advances that are being achieved. Advances in 3D printing, additive manufacturing and other new production systems are not only resulting in spacecraft operating more efficiently and cost effectively, these new techniques are also serving to reduce the cost of launch vehicles as well. The advances of today will not be the advances of tomorrow, though. The space revolution will continue in the form of new and better forms of space transport.

In the 2020s and 2030s even more innovative systems will provide dramatic improvements in our ability to get systems to orbit. Deep space transportation will improve the most, as we find even more efficient ways to propel systems into the cosmos. We have found that Earth orbit and deep space systems diverge in that those systems needed to climb out of Earth's gravity well need to build up velocity via constant acceleration over weeks, months or even years for deep space missions.

Deep Space Systems

There are many ideas about how to create new and innovative systems optimized for deep space travel. These systems would include those that have the ability to fly to other stars and to leave the Solar System on interstellar flights. There are at least four basic concepts that have been seriously considered that could attain velocities sufficient to travel outside the Solar System. These concepts involve: (i) solar sail systems that could also be augmented by gravity assist; (ii) laser driven or directed energy systems that likewise could be aided by gravity

J. N. Pelton, *Space 2.0*, Astronautical Engineering,
https://doi.org/10.1007/978-3-030-15281-9_11

assist; (iii) nuclear fusion-powered vehicles; and (iv) systems that would combine water-ice fuel systems with protective shielding against collisions with particles while traveling at substantial velocities near light speeds.

Solar Sail-Powered Craft

The idea of solar sail-powered craft is hardly a new idea. In recent years, though, there been practical experiments seeking to determine the possibility of using such propulsion systems to reach sub-light speeds that would be needed to travel to other stars. The Planetary Society has attempted several experiments with its "LightSail" approach that deployed small objects from a nanosat. These experiments to date have had limited success.

The Japanese agency (JAXA) with its IKAROS experiment in 2010 was the first to prove that solar-powered light sails indeed could work. This light sail was composed of a polyimide with a reflective surface and was 200 square meters in size. It traveled to Venus and then traveled on to the far side of the Sun.

The European Space Agency had their Gossamer De-Orbit Sail experiment in 2013. Currently NASA has its Near Earth Asteroid Scout that is scheduled to fly on the Solar Launch System (in 2019). JAXA has its OKEANOS experiment that will seek to fly to Jupiter's Trojan asteroids in the late 2020s (https://en.wikipedia.org/wiki/Solar_sail#Breakthrough_Starshot).

The most exotic and ambitious project of all is known as Breakthrough Starshot which is privately funded. This light sail experiment, which is headed by ex-NASA Ames Chief Gen. Peter Worden, would seek to achieve an incredible speed of 20% of the speed of light (60,000 kilometers per second or about 134,000,000 miles an hour). In short, this fantastic spacecraft would be gradually accelerated until it would be going really, really fast.

This amazing craft would be composed of lightweight elements for its light sail. It would not be driven by solar radiation but instead will be powered by a number of Earth-based lasers that would drive a nanocraft carrying miniature cameras. This 'Starshot' project would seek to go to Alpha Centauri, the closest star to Earth, which is some 4.4 light years away (about 40 trillion kilometers distance). That is equivalent to about 27,000 trips to the Sun – or a very long, long way.

There have been many different materials proposed for the structure of a solar sail vehicle (i.e., lithium or polyimide), for an effective reflective surface (i.e., aluminum) and for a good emissive surface (i.e., chromium). There are variations on the theme of light sails. One option is also the use of the solar wind (or particles emitted from the sun to create a magnetic field) or a magnetic sail that could be used to generate power to support electrical propulsion. One can also embed high-efficiency solar cells in the light sail to generate electricity for ion propulsion. Finally it might be possible to use the gravitational field of perhaps Jupiter or Saturn to slingshot a light sail craft to a higher velocity.

Today the feasibility of light sail systems for scientific or commercial purposes remains to be proven. The continuing efforts by NASA, ESA and JAXA to test light sails together with the private initiatives of the Planetary

Society and the ambitious Breakthrough Starshot initiative suggest a high level of interest in a light sail starship. If you believe that all of these space scientists in their mission to make light sails work are not bonkers, then one of these projects may truly succeed. One can at least hope that they are not squandering their efforts by chasing bogus photons – the modern version of tilting at windmills. All of this effort seems to suggest that there is truly something there. We must at least hope that one of these efforts will eventually be proven to be correct and true starships created. The physics and the theory are clearly there already. Perhaps we will find the way to create workable light sail starships within the next decade or so. Figure 11.1 shows how laser signals can be used to drive a human-designed craft up to the highest velocity ever known.

Laser or Ion Driven Spacecraft

The light sail systems envisioned to date would use either photons from the Sun or an Earth-based laser system. Some scientists have envisioned that photons or electrically charged ions generated on board a spacecraft can achieve great speeds over time. This type of system would not need to depend on the Sun or laser or directed-energy systems situated on Earth, but rather could travel with the spacecraft to accelerate it to ever greater velocities. Again the

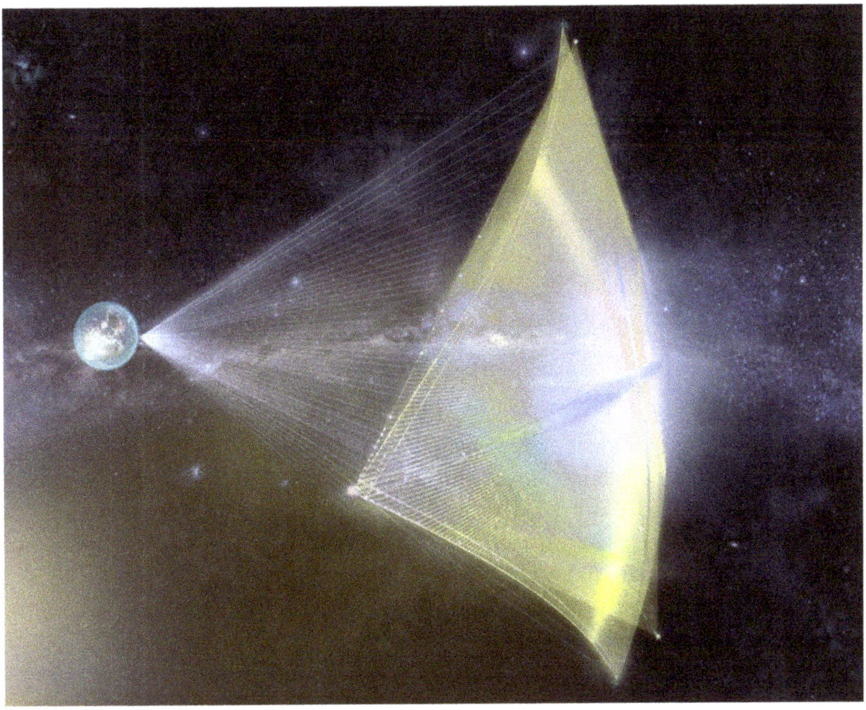

Fig. 11.1 Artist representation of Earth-based laser system connecting with a starship. (Graphic courtesy of Breakthrough Starshot.)

concept is not that of great bursts of speed but the steady accumulation of velocity through constant acceleration for long periods of time. If such a system could only achieve a tiny acceleration of 1 cm/sec^2 it would nevertheless add up over time. The acceleration of gravity at ground level is 9.8 m/sec^2, and thus this acceleration would be 980, or nearly a thousand times less than the force of gravity on our planet's surface. Over time this modest acceleration would add up. There are 86,600 seconds in a day and 31,586,600 seconds in a year. Thus the distance traveled in a year would be about 50 million kilometers – not even half way to the Sun. But in two years it would have gone 200 million kilometers, in three years 800 million kilometers, in four years 1.6 billion kilometers, and by the end of five years about 3.2 billion kilometers. The squared part of the formula relating to acceleration adds up over time.

The key would be to design an on-board system that can last for many years, one that can generate electrically charged ions or emit photons for many years at a time so that the velocity can build up to a speed sufficient to travel to the stars. The Breakthrough Starshot initiative has based its mission design on using a very high-powered ground-based laser, but there is no particular rule that says that a combination of technologies could not be used. It might be possible to use an interesting combination of thrust systems. One might use a ground-based laser, an on-board laser system, an on-board ion thruster or even a chemical thrust system to maneuver around the orbit of Jupiter or the Sun to create a gravity-assisted boost to a spacecraft's velocity that might be the most efficient. Indeed it might be hybrid systems that use a combination of thrust technologies that prove the most effective way to achieve the maximum speeds and still have the maneuverability that is needed to avoid obstacles and use gravity assist as part of the process.

Nuclear-Powered Vehicles

The U. S. Defense Advanced Research Projects Agency (DARPA) and the British Interplanetary Society have been considering the feasibility of nuclear-powered transportation systems for a half century. It has been suggested that nuclear fission or even nuclear fusion systems have sufficient concentrated power to allow for efficient long-range space travel. There are today many feasible and efficient spaceship designs for different ways to use nuclear energy for propulsion.

Although we are not expecting to create starships such as in *Star Wars* or *Star Trek*, with craft that can make hyperspace leaps from one part of the cosmos to another and exceed the velocity of light speed limit, there is still enormous potential for a nuclear-powered craft that might be continuously accelerated until it did reach enormous speeds.

The first steps in the use of nuclear energy in space were to use isotope-based and plutonium-based power supplies for long-term operation of spacecraft. The so-called Radioactive Thermo-electric Generators (RTGs) have been proven on spacecraft for many years. In the case of RTGs they simply convert heat from plutonium-238 decay into electricity, using thermocouples. The RTG on the Navy's Transit 4A satellite produced only about 3 watts of electrical power. In the late 1960s, NASA launched the RTG-powered Nimbus III, which lasted decades in orbit due to its nuclear power supply.

There has been continuing research and development in the area of rocket propulsion that goes beyond simply powering a spacecraft in its operation. There are several options that are subject to current R & D. They fall into categories such as: (i) nuclear pulse propulsion; (ii) nuclear thermal rocket propulsion; (iii) nuclear electric propulsion; (iv) direct nuclear propulsion; and (v) nuclear-powered ramjet propulsion.

Nuclear Pulse Propulsion This type of approach to nuclear propulsion traces its history to the Orion (DARPA) and Daedalus (British Interplanetary Society) research projects back in the 1960s and 1970s. The most recent concepts for a starship designed with nuclear pulse propulsion involve what is called catalyzed antimatter. This is a variation of nuclear pulse propulsion based upon the injection of antimatter into a mass of nuclear fuel as a means of reducing the size of the nuclear pulse propulsion system.

Nuclear Thermal Rockets This is another type of nuclear fission reaction propulsion system and can be used to create propulsion in a variety of ways. It would typically not be used for liftoff from Earth, but for efficient station-keeping operations that would be more efficient than chemical propellants. Nuclear thermal rockets can provide great performance advantages compared to chemical propulsion systems. Nuclear power sources could also be used to provide the spacecraft with electrical power for electrical ion thrust.

Direct Nuclear Propulsion Again there are many variations on this basic concept. One of the prime concepts would involve a core reactor that directly propels the rocket by the exhausted coolant in a gaseous fission reactor. The nuclear fission reactor core could be either a gas or plasma. This type of direct nuclear propulsion could possibly create specific impulses that would be in the range of 30 to 50 kilo Newton-seconds/kg of fuel. It would be based on exhaust velocities that could be as high as 50 kilometers/second. This exhaust plasma or gas would create sufficient thrust for fast planetary or even interplanetary travel. In some designs hydrogen coolant becomes the propellant. The hydrogen serves to cool the reactor and its various structural parts. Once the hydrogen passes through the core region, it is exhausted at the indicated velocities. If cooling from the propellant is not enough, external radiators can be used for the reactor as a supplement. In some other designs, fissioning gas plasmas are used to heat a low mass propellant.

Nuclear Ramjet Propulsion This is the approach to use nuclear power to create a ramjet approach to a cruise missile or aircraft. Project Pluto is one such development project. This type of system only works inside Earth's atmosphere. The principle behind the nuclear ramjet is actually quite simple. The velocity of the missile sucks air into intake vents at the front of the vehicle. This is known as the ramjet. The innovation is to use a nuclear reactor to superheat the compressed air that has already experienced the ramjet effect. The super-heated air exhausts at high speed out through a nozzle to create thrust sufficient to result in hypersonic speeds.

In the future there are many potential applications of nuclear fission or fusion-based space transportation systems. Propulsion systems that allow for efficient station-keeping operations using electric ion propulsion systems that outperform today's xenon ion propulsion systems in terms of long-life performance are the most likely near-term application. Today's chemical propulsion systems are quite adequate to get to the Moon and the closer planets, and gravity assist can help to get to even the outer planets.

Even so NASA has at least argued that a nuclear-powered rocket could be half the size of a chemically powered rocket to support missions to Mars. Figure 11.2 shows a nuclear thermal propulsion system that was proposed to the U. S. Congress in 2016. At the time, NASA administrator Charles Bolton proposed that it could create such nuclear fission-based craft by 2033. They provided testimony as to why it could halve the size of a chemically powered rocket system. Such a program has not been authorized for funding [1].

At this time it seems that true nuclear rocket systems will have to await future space missions. Only such objectives as planetary defense against potentially hazardous asteroids, Mars colonization, or asteroid mining might convince legislative bodies to fund such a difficult and expensive program. This is in spite of the cost efficiencies that have come from NewSpace commercial programs that have seemingly produced far more effective technology at lower cost in recent years.

The bottom line is that it seems that there will need to be more powerful economic incentives to develop either direct nuclear propulsion or nuclear pulse propulsion systems in coming years by NASA or any other space agencies. There is always, of course, the latent goal of developing nuclear-powered rocket systems for objectives related to

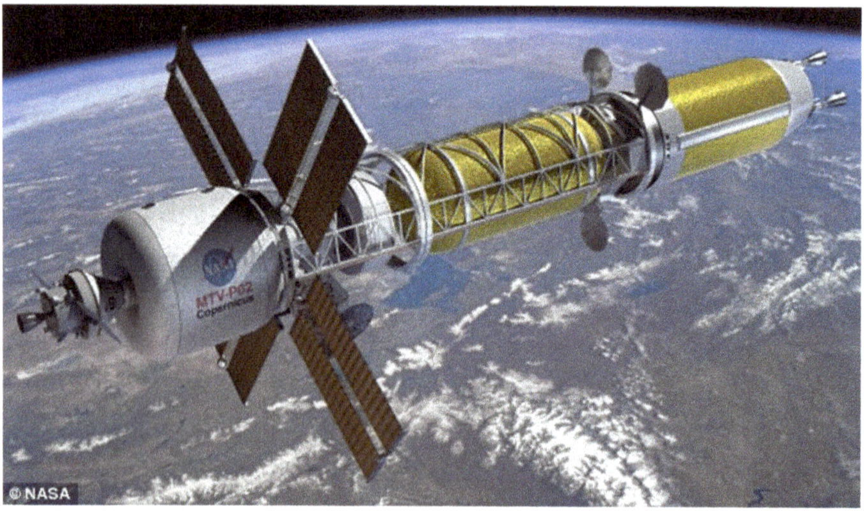

Fig. 11.2 Proposed nuclear thermal propulsion craft called Copernicus, envisioned by NASA to go to Mars. (Illustration courtesy of NASA.)

national defense. Such a step would be a step backwards for commercial development of space and the peaceful uses of outer space.

H_2-O_2 Defined Star Craft

Most studies of starships using nuclear propulsion focus on the most pertinent issue of how to achieve sufficient thrust to generate velocities that represent a significant fraction of the speed of light, especially when it is understood that Lorentz contraction also means that mass increases as it moves to higher and higher speeds. But even if we humans can design interplanetary craft that can move at speeds such as a third of the speed of light, there will be other problems to be addressed and solved. One of these key issues is how to protect a star craft that is moving at such enormous speeds. If such a spacecraft should encounter a substance such as ice formation in the Oort Cloud moving at enormous relativist speed the result would likely be disastrous. A magnetic shield might help to divert ions, but some form of physical shield or buffer would seem to be required for objects that hold no electromagnetic charge.

Brian McConnell in his book about water-based "spacecoaches" has made a credible technical case for a starship that is essentially made of frozen water as its essential makeup, with an ice-shield against cosmic accidents. This is an idea perhaps first put forward by Arthur C. Clarke, in which a star cruiser has to land to restock on ice. McConnell has stated his arguments for his water-based design as follows: "The water is used for many purposes before it is superheated via electro-thermal engines to generate thrust, uses which include radiation shielding, thermal management, life support, crew consumption and attitude control." The most important use of all, however, might be a protective ice shield that would protect the pursuing spacecoach [2].

Tethers, Space Elevators and Space Funiculars

The nearer term future may look to the more expanded use of tethers and then, over the longer term, ultimately of space elevators or funiculars. The use of tethers or long extended structures in Earth orbit can be used both to lift objects to higher orbits and also to generate electricity within the geomagnetosphere to create the energy needed to restore the tether lifting system after the transfer orbit has been achieved. Several experiments by NASA, JAXA and ESA have demonstrated the feasibility of momentum exchange to lift objects to higher orbits.

NASA is currently working on an experiment known as the Momentum-Exchange Electrodynamic Reboost tether propulsion system, or MXER tether. The object of this experiment is to deploy a very long tether, some 100 to 150 kilometers in length, that would be able to go through a momentum exchange to lift a satellite from low Earth orbit to a transfer orbit that that would have an apogee at GEO orbit. The NASA explanation of how this process would work is as follows:

> *Momentum-exchange tether propulsion transfers momentum from one object to another by briefly linking a slow-moving object with a faster one. Much the same way as ice skaters play "crack the*

whip," the slower object's speed could be dramatically increased as momentum and energy is transferred to it from the faster object. Similarly, a spinning tether facility in an elliptical Earth orbit might snare slower-moving spacecraft in low-Earth orbit and throw them into much higher-energy orbits [3].

This type of momentum exchange lifting system, if proven in practice, could significantly reduce the cost of launching satellites to GEO orbit and would require large communications satellites to be launched to LEO, where this tether lift system would raise them to a transfer orbit before final deployment in the correct circular orbit.

Previous experiments have involved tethers of much shorter lengths. The MXER experiment, however, would require a much greater length to generate enough electrical energy to restore the system to its original orbit so that it could repeat the lift sequence over and over again.

The idea of using Earth's magnetic field to generate electricity is not only foreseen as a way to lift satellites to higher orbits. The so-called EDDE systems (Electrodynamic Debris Eliminator) envision a system that is used to throw a net over space debris to hasten its de-orbit and operate in orbit over the longer term by generating electric propulsion derived from Earth's magnetic field as well [4] (Fig. 11.3).

An in-orbit momentum exchange lift system that would be up to 150 kilometers (94 miles) in length represents the most ambitious tether experiment to date. Yet this is far, far short of the idea of creating a space elevator or space funicular to GEO orbit. In this case the

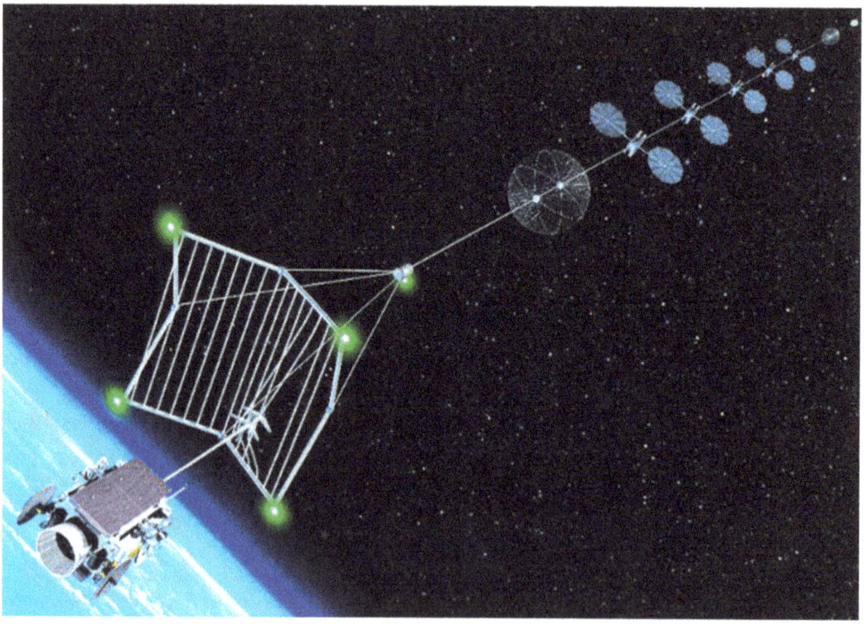

Fig. 11.3 Artist representation of the Momentum Exchange Lift System (MXER) experiment by NASA. (Graphic courtesy of NASA.)

tether or cabling system would not only need to reach from Earth's surface out to GEO orbit but considerably beyond to create a lift capacity – or 'negative weight' – that would offset the gravitational pull of a giant tether system that reaches 35,870 kilometers (22,230 miles) out to GEO orbit. From this perspective, the momentum exchange system that would lift a LEO satellite to a transfer orbit, then to GEO seems a much more feasible and cost-effective solution.

Mass Drivers and Rail Guns

Other advanced space transportation systems that have been considered for some time are mass drivers or rail guns. Again the key force involved is electromagnetism. There have been proposals put forth, from Maglev trains to the HyperLoop and rail guns, for years.

The basic idea with rail guns or coil guns is to use magnetic force to accelerate transport vehicles or units to high rates of speed. Gerard O'Neil in his book *The High Frontier* explained how a Maglev system on the Moon could send excavated materials from the Moon into orbit with relative ease due to the Moon's low gravity and lack of an atmosphere. The materials sent could be used to supply basic building materials to create a space colony. Creating a Maglev system that would attain sufficient speed to gain orbit from Earth is a much more difficult feat.

Rail guns or coil guns create an electromagnetically induced acceleration along its rails (or coils). If the projectile or launching device (i.e., sliding armature) is magnetically active the speed will increase exponentially as power is increased. If the rails are many kilometers long, very high velocities can be reached – especially in a vacuum where there is no resistance.

The practical aspects of coping with the problem of g-forces and atmospheric drag currently suggest that a rail gun or coil gun would be used for launching materials or cargo into space rather than subjecting humans to excessively high g-forces or trying to cope with atmospheric drag. Another option would be to design a rail gun or coil gun system that might be used to assist with a spaceship launch. It has been suggested that one might create a rail gun that could be operated from a very high mountain or elevated structure that might be able to launch a stream of pellets or even propel a vehicle into space [5] (See Fig. 11.4).

Current understanding of the technical constraints suggest that human launch from Earth would not be possible. Use of this concept to launch humans from the Moon, Mars or other planetary bodies with limited gravitational force and little or no atmosphere, however, could be possible.

A rail gun or coil gun could, of course, be used as a weapons system. In this case, the objective would be to create very high speed targeted projectiles rather than putting cargo into space. Tests of a rail gun weapons system with energy levels that involve many mega Joules of power have been tested, but only in experimental tests. Such projective firings would not be to support launch operations but as longer range destructive cannon-type systems.

There is also the idea of accelerating plasma via a helical rail gun design with sufficient temperature to even achieve and sustain nuclear fusion. This is one of the concepts that is being explored to create a sustained nuclear fusion process. This sort of idea, however, has to be considered as a much longer-term

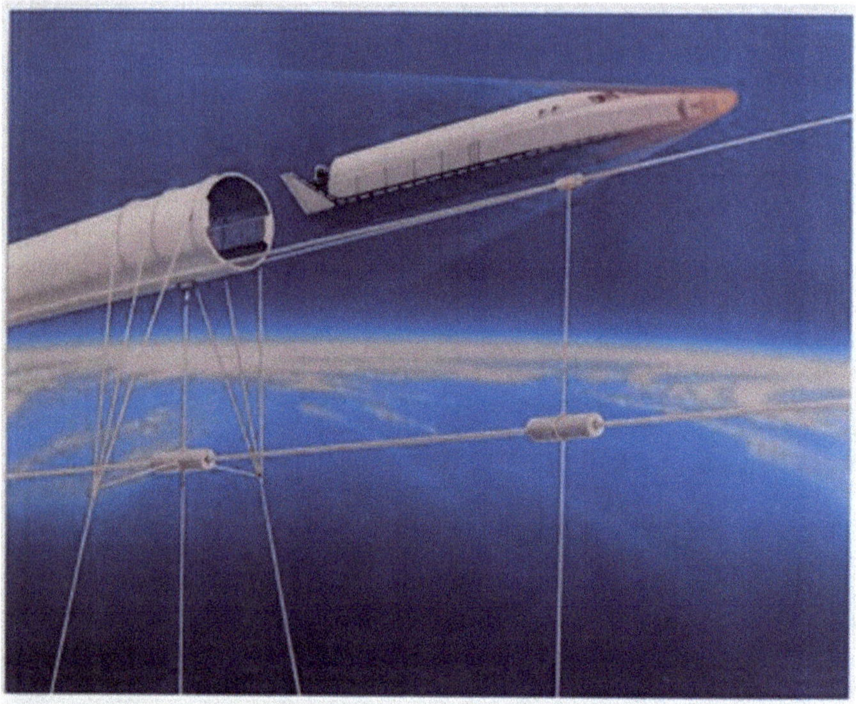

Fig. 11.4 Artist concept of a coil gun or Maglev system that could assist with the launch of a spaceship (Illustration courtesy of NASA.)

research activity and not a near-term concept.

Space Shields and Large-Scale Construction Projects in Space

For many years, the idea that one might create large-scale construction of infrastructure in outer space has been considered science fiction for several legitimate reasons – extremely high cost; technical, scientific and engineering difficulty; extremely harsh space environment; higher priorities here on Earth; and lack of an international regulatory framework or global space governance process under which to undertake such an initiative.

Today the rate of scientific and technological progress in space systems is hitting new heights. The sky is no longer the limit. Soon the James Webb Telescope will be launched, and we will be able to see back virtually to the Big Bang. The International Space Station (ISS) has set a precedent for international cooperation by dozens of countries working together to create and operate a complex facility in space to advance scientific knowledge and foster global cooperation in outer space. The opportunity now exists to create new space systems to protect our world from devastating cosmic hazards.

There are new types of infrared telescopes that could detect potentially hazardous asteroids down to 30 meters in size (city killers) as opposed to systems that can spot hazardous asteroids only 140 meters and larger. There are new systems that might allow us to create magnetic shields that could protect us from devastating coronal mass ejections from the Sun that could wipe out our electric power grids, our pipelines and vital space systems. We now have the technology and the knowledge that could protect our planet from devastating 'black swan' events that could do trillions of dollars of harm to the world economy. New types of technologies can allow us easier and more cost-effective access to space with new ways to protect the global economy from devastating cosmic hazards and preventing huge loss of life. It has been assumed for many years that we can only suffer the losses that come from killer asteroids, solar storms that destroy our vital infrastructure and cosmic hazards. Today we are developing new space technologies that show us a pathway to safeguarding Earth in new and innovative ways.

Conclusions

The majesty of launch vehicles propelling spacecraft into space is an awesome and exciting sight. New launch systems are becoming more reliable, effective and cost-efficient. Yet the subjects discussed in this chapter suggest there are important new technologies in the pipeline that could be safer and more efficient. The idea that the best way to put people into space by putting them on top of a controlled bomb will be likely become passé within the next three decades. It is not only dangerous, but there are environmental hazards that come from spewing noxious chemicals from more and more rockets launched at an ever increasing pace.

We need to find ways to do more in space even more efficiently. We need to build new space infrastructure to create clean electrical energy systems, protect our planet and provide better space-based services. We can accomplish this by developing the latest and best new means to get stuff into orbit. We still have a long ways to go to find and perfect the best ways to do this in the most efficient ways possible. If we can use tethers to flip spacecraft from LEO to GEO, the environment and the space industry will benefit. Nuclear propulsion, electric ion propulsion, solar sails, rail guns, and perhaps ultimately space elevators will create a new and even more exciting new space age in the decades ahead. The short story is that the coming trillion-dollar space industry will be built on the base of a host of new space technologies and exciting new ways to get spacecraft into Earth orbit.

References

1. Zolfagharifard, E.: NASA wants to use nuclear rockets to get to Mars: space agency claims the technique is 'most effective way' of reaching red planet. Dailymail.com (2016). http://www.dailymail.co.uk/sciencetech/article-3499441/Nasa-wants-use-nuclear-rockets-Mars-Space-agency-claims-technique-effective-way-reaching-red-planet.html
2. McConnell, B.: A Design for a Reusable Water-Based Spacecraft Known as the Spacecoach. Springer, Cham (2015)
3. NASA: momentum exchange tether propulsion. https://www.nasa.gov/centers/marshall/pdf/100415main_momentum.pdf. Accessed 30 July 2018
4. ElectroDynamic debris eliminator. http://www.star-tech-inc.com/id121.html. Accessed 30 July 2018
5. Atkinson, N.: NASA Considering Rail Gun Launch System to the Stars. Universe Today (2010)

12 Spaceplanes, Space Tourism and Private Space Habitats

Introduction

Spaceplanes carrying celebrities, sports figures, movie stars and royalty will soon be the rage for the next few years – barring a serious accident. Sir Richard Branson has done well to book not only millionaires but media idols to promote his space adventures business known as Virgin Galactic. His VSS Unity spaceplane is now set to carry would-be citizen astronauts up 120 kilometers into space. Jeff Bezos, with his increasingly successful Blue Origin suborbital flights will apparently soon follow suit. In what is not always friendly rivalry with Musk and Branson, Bezos's company will also be booking big names to capture headlines and promote his suborbital launch service as well.

At the start, both Branson's and Bezos's companies will provide customers with about four minutes of weightlessness and a chance to see the big blue marble we call Earth in the dark sky of outer space during a several-hours-long flight.

Robert Bigelow, whose fortune is based on the Budget Suites hotel chain, heads Bigelow Aerospace, and his goal is commercial space habitats. His innovative company is pursuing the operation of inflatable space habitats for those willing to pay for a true trip to space and a longer stay. The key question is whether the space tourism business, which now finally seems to be on the point of blossoming into paying services, represents an economic bonanza or simply headline hype?

Space tourism services are perhaps just a small part of the overall Space 2.0 enterprise, but nevertheless they can play a key role. That role is to keep NewSpace companies in the news and firing the imagination of a global public that has perhaps grown weary of space agency accomplishments that seem to come at tremendous expense.

As we have seen in earlier chapters, the space industry is currently closing in on becoming a $400-billion enterprise and seems headed toward becoming a trillion-dollar business in the decade ahead. But it is things like Elon Musk's boldness in launching his Tesla into orbit, or Richard Branson launching A-list celebs into orbit, that captures newspaper ink or global television news. The truth is

J. N. Pelton, *Space 2.0*, Astronautical Engineering,
https://doi.org/10.1007/978-3-030-15281-9_12

that actual space tourism businesses, according to detailed market studies, are not expected be a large percentage of the Space 2.0 total market. The leaders of the space tourism industry themselves are changing their business models to capture additional revenue streams.

Certainly Richard Branson and Virgin Galactic have altered their business plans to develop a small satellite launcher, called Launcher One. This vehicle is clearly designed to augment revenues. The contract to use Launcher One for several dozen OneWeb satellites now represents a significant portion of the projected future revenues of Virgin Galactic.

Robert Bigelow, of Bigelow Aerospace, has indicated that his space habitats will also be available for low-g experiments. Jeff Bezos's Blue Origin is intent on developing full launch capacity to provide commercial launch services that go well beyond space tourism services. Swiss Space Systems (S-3), now bankrupt, was developing a spaceplane capability, but also announced detailed plans about how it intended to use its spaceplane not only for suborbital flights but also as an intermediate lift stage that would allow a final stage launcher to lift small satellites to orbit. Indeed Virgin Galactic has implemented a similar strategy by developing its Launcher One small satellite launcher to supplement revenues from its suborbital flights for its space adventures.

This chapter thus not only explores the development of the space tourism industry to date, but also examines where new space enterprises in this sector of the Space 2.0 industry is now headed. Are these various efforts to develop spaceplanes and private space habitats aimed well beyond just the space tourism market? Are they spearheading a number of new ventures that ultimately will open up major new markets? Do these new markets include such new enterprises as hypersonic transportation systems and other space transport systems? Are carrier vehicles and spaceplanes just the first step towards innovative new ways to launch spacecraft into orbit? Is Sierra Nevada's Dreamchaser spaceplane just as key to the future as Branson's SpaceShipTwo? And finally, has the importance of Bigelow Aerospace Genesis 1 and 2 in orbit more to do with new and more cost-effective ways to carry out micro-gravity experimentation than to do with space tourism?

The answers to these questions are still far from clear. It may turn out in the strange and wonderful world of Space 2.0 that the final answer might be all of above. The interesting thing is that some very clever space entrepreneurs are keeping as many options open as possible. Space billionaires Elon Musk, Richard Branson, Jeff Bezos, Robert Bigelow, Paul Allen and others are doing more than developing a future path to more exciting and cost-effective space tourism. No, they are about much more. They are truly opening the door to a wide range of new Space 2.0 business opportunities as well.

The XPRIZE and Efforts to Build Spaceplanes to Carry Citizen-Astronauts into Space

The person who was one of the first to recognize the serious potential of what we now call NewSpace or Space 2.0 was the author's friend and colleague Peter Diamandis. When we worked to set up the International Space University (ISU) in the mid-1980s, some 30 years ago, it was Peter who conceived of the iconic poster that showed a university in space. This remarkable artwork fired the imagination of the first hundred students to attend the sessions at MIT. It was Peter who conceived of the $10 million XPRIZE to fuel the competition to create the world's first privately funded development of a spaceplane.

At the time the set objectives seemed impossible to achieve. The goal was for a pilot and crew member to fly a spaceplane up above 100 km and then descend to land safely. Then, they had to do it all over again within an 8-day period.

It was a challenging feat indeed. The objective at the time seemed to be so unlikely at the time that an insurance company provided the Ansari family a $10-million policy against this happening. It turns out that the Ansari family, who underwrote the prize in this manner for a $1 million outlay, were shrewder than the hapless insurance company. The insurers had to pay out when Microsoft co-founder Paul Allen and aerospace designer Burt Rutan designed SpaceShipOne and the carrier plane White Knight that twice achieved this incredible feat. When pilot Mike Melvill successfully landed in the Mojave Desert for the second time on October 4, 2004, an impromptu sign was hoisted aloft that read: "SpaceShipOne-Government Zero." [1] This bold initiative that combined the wealth and moxie of computer entrepreneur Paul Allen and the unconventional aerospace design of Burt Rutan signaled a new day in the commercial space industry.

The NewSpace initiatives that followed frequently combined the inventive genius of Silicon Valley and the thinking of traditional aerospace industries. The next major initiative of Peter Diamandis is indicative of the new Silicon Valley focus that is seen over and over again in Space 2.0 enterprises. Peter worked with A. I. guru Ray Kurzweil, who gave us SIRI, and Pete Worden, then director of NASA Ames and now head of the amazing Breakthrough Starshot initiative, to start the Singularity University. This counterpart to the International Space University is seeking to train young entrepreneurs to start new ventures – often Space 2.0 start-ups – to make a positive impact on the lives of over a million people. The SU venture in Mountain View, California, on the grounds of NASA Ames and in the heart of Silicon Valley, is striving to bring new thought and innovation to the world by training young entrepreneurs from all over.

New Models of How to Get to Space

Based on the successful test flights of the SpaceShipTwo VSS Unity spaceplane in July 2018, it is expected that actual commercial flights to bring the first 500 citizen-astronauts on suborbital flights to outer space will be the featured Space 2.0 accomplishment of the next few years (Fig. 12.1) [2].

As of July 2018, the Virgin Spaceplane System (VSS) Unity had flown for the first time successfully into the stratosphere. This successful test and fault free landing at the Mojave Desert spaceport facilities was yet another indication of a successful recovery from the disastrous Halloween crash of the

Fig. 12.1 Landing of SpaceShipOne, which that claimed the $10 million Ansari XPRIZE. (Graphic courtesy of Virgin Galactic.)

Fig. 12.2 Free flight of the VSS Unity spaceplane into the stratosphere after separation from carrier plane Eve. (Graphic courtesy of Virgin Galactic.)

earlier version of SpaceShipTwo that occurred in October 2014 [3].

Currently there are still some 640 would-be citizen-astronauts who have signed up at either $200,000 or $250,000 for a short ride into space on SpaceShipTwo. And Jeff Bezos's test of his vehicle with its vertical takeoff and capsule landing design is not far behind (Fig. 12.2).

SpaceShipTwo takes off horizontally while hoisted on a carrier aircraft and also lands like a conventional aircraft on a runway. Jeff Bezos's Blue Origin Company is very much taking the opposite approach to carry citizen-astronauts into space. His New Shepard vehicle takes off vertically from a launch gantry and then the passengers and crew come back floating down inside of a capsule

Fig. 12.3 The New Shepard vehicle with return capsule during a July 2018 test flight. (Graphic courtesy of Blue Origin.)

Table 12.1 Comparison of energy required for different types of vehicles.

Comparing airplanes, jets, and spaceplanes to rockets to orbit				
Comparison	Airplane	Jet	Spaceplane	Rocket to LEO orbit
Velocity (m/s)	250	500	1600	7800
Height (km)	Up to 10	Up to 20	Up to 120	200+
Specific energy (Joules/kg)	0.13	0.7	14.5	324

on a parachute. The New Shepard design is heavily focused on providing crew and passengers an escape option at every phase of the operation to maximize safety (Fig. 12.3).

Blue Origin has aspirations that go well beyond providing rides to space tourists on suborbital flights. Its New Glenn vehicle is intended to be able to fly into space. But certainly they are difficult transitions to make the climb from jet plane to suborbital spaceplane to launch vehicle capable of delivering spacecraft to orbit. The following chart shows how challenging such transitions are in terms of the energy required that (expressed in Joules/kg) goes up exponentially rather than linearly (Table 12.1) [4].

Virgin Galactic and Blue Origin are not alone in their efforts to create commercial capabilities either to provide suborbital spaceplanes or launch satellites to orbit. As previously noted, Boeing and SpaceX are also seeking to provide commercial launches to the International Space Station (ISS) for NASA under commercial crew contracts, and Sierra Nevada continues with the development of their Dreamchaser spaceplane, which can be launched to orbit via vehicles such as the Atlas V and then bring cargo and potentially crew back from low Earth orbit (LEO).

And this is just U. S.-related commercial programs. ESA is working with industry in Europe to develop spaceplanes and commercial launch capabilities. There have been various initiatives with Airbus Defence and Space, Bristol Space Planes and other contractors to develop spaceplanes under the Future European Space Transportation Investigations program and its follow on, the so-called PRIDE program [5].

Efforts now include those of Reaction Engines to develop the Sabre engine and the Skylon vehicle to prove the viability of a ramjet engine propulsion system working in tandem with a rocket engine to allow an efficient and reusable single-stage-to-orbit vehicle. This is a hybrid rocket system and ramjet spaceplane that developer Alan Bond has said will revolutionize the space launcher industry [6].

Indeed there are active research programs to develop spaceplane technologies and systems in Japan, India, China and Russia at various stages of capability. Most of these, however, are at the governmental level rather than as privately funded capital ventures.

Many of the initiatives to develop spaceplanes started off with the XPRIZE competition, and this gave rise to an incredible diversity of concepts as to how these systems would take off and

Table 12.2 New approaches to space inspired by the XPRIZE competition. (Prepared by the author and derived from a chart created for the International Space University.)

Efforts to create new commercial systems for space tourism in the past 15 years	
Various approaches for accessing space	**Companies using this particular approach**
Lighter than air ascender vehicles and ion engines with high altitude lift systems providing access to LEO	JP aerospace
High altitude experience from stratospheric dirigible ascent	World view, zero-to-infinity
Balloon-launched rockets with capsule return to ocean by parachute	Planetspace
Launch space plane to orbit on conventional launcher and horizontal landing	Sierra Nevada Dreamchaser
Vertical takeoff and vertical landing	Armadillo aerospace, blue origin, Lockheed Martin, Masten space plus new SpaceX grasshopper project
Vertical takeoff and horizontal landing (spaceport)	Aera space Tours, Planetspace, SpaceDev., SpaceX, Sub-Orbital Corp, t/Space, TGV rocket, Wickman spacecraft & propulsion
Vertical takeoff and horizontal landing (from ocean site)	Advent launch site
Horizontal takeoff and horizontal landing	Andrews, scaled composites, the spaceship corporation, virgin galactic, XCOR
Tow launch and horizontal landing	Kelly Space & Technology, Inc.
Vertical launch to LEO from spaceport	Alliant ATK (now orbital ATK), Inter Orbital Technologies, SpaceHab, UP aerospace
Launch to LEO from carrier jet drop	Triton systems, Stratolauncher, launcher one

land. Table 12.2 provides a summary of just the U. S. XPRIZE-inspired initiatives to develop a better and more cost-effective approach to creating a spaceplane business. Some might suggest that this was an incredibly inefficient process. But such a competitive thought process, largely inspired by the XPRIZE competition, has weeded out weaker ideas and business practices and has left the strongest technical approaches now moving ahead [7].

Today there are still many ideas moving ahead, although a majority of the XPRIZE initiatives have now folded. The net result is that there is more coherence and focus on the approaches defined by Virgin Galactic and Blue Origin in terms of space tourism flights, although one could also say that the stratospheric flights by dirigibles, as being pursued by World View and Zero-to-Infinity, were also originally inspired by the XPRIZE competition.

Private Habitats in Space

Robert Bigelow, who created the Budget Suites of America string of hotels, has always had a desire to be more than just another Las Vegas-based hotelier. His reputation among the hotel and casino owners of Las Vegas is that he was a loner, and when he founded his aerospace company in 1998, it was clear he was not just your usual hotel magnate and entrepreneur.

Bigelow went on to purchase the license from NASA to develop inflatable space habitats that had been originally created by SpaceHab under U. S. government funding. He has launched two Genesis space habitats that are now fully deployed in space, with cameras streaming down video from their low Earth orbit. His next step was to design a much large inflatable habitat called BEAM. This stands for Bigelow Expandable Activity Module. Under contract with NASA he has deployed a prototype on the ISS to create expanded living and experimental space for ISS crew members. Although there was an initial problem with this unit, it has now successfully deployed. See the BEAM prototype as deployed on the ISS; it's the ovoid structure at the upper middle of Fig. 12.4 [8]. This BEAM structure could be made much larger.

Bigelow has ambitious plans that go well beyond the BEAM experiment on board the ISS. He feels he has the key parts of the design of the habitat with a viable life-support system and a solar array power system well in hand. The key missing element was a reliable and cost-effective transportation system to and from a private space station. In this respect he is seeking a U. S. developer of a rocket launcher system that could ferry experiments and space tourists for a stay on his private space habitat.

How could he find a reliable way of transporting humans to and from low Earth orbit (LEO)? In 2004 Bigelow launched his own competition and promised the winner a $50-million payout. This competition he dubbed America's Space Prize, but its conditions of performance and commitment to provide private ferrying services to Bigelow made it a very long shot that there would be any viable competitors. In early 2010, over five years later, the prize offer expired without a winner.

In fact, in August 2009, Bigelow Aerospace essentially abandoned the competition initiative by announcing an effort to develop the so-called Orion

Fig. 12.4 BEAM Inflatable barrel-like structure as deployed on the Tranquility node of the ISS. (Graphic courtesy of NASA.)

Fig. 12.5 Private space station concept for Bigelow private space habitat. (Graphic courtesy of Bigelow Aerospace.)

Lite spacecraft. The idea was to use the Orion Lite capsule in tandem with either an Atlas 5 or Falcon 9 launch system. This system would be able to carry a 'pilot' and up to six passengers [9].

Bigelow has provided a mockup of his plans for a module-like space station as shown in Fig. 12.5. He has announced fees for not only space tourist visits to his private space station but also for in-orbit low-g experiments. There have also been early indications that governments and various pharmaceutical, chemical and biological industries are likely to sign up for these types of in-orbit activity.

Nor is Bigelow Aerospace the only entity that has contemplated creating private space habitats or space hotels. German aerospace engineers such as Krafft Ehricke considered the technical aspects of such a design, and von Braun worked with Walt Disney on such a concept. Stanley Kubrick and Arthur C. Clarke created a vivid image of what a Hilton hotel in space might be like in *2001: A Space Odyssey*. More recently, the Spanish space hotel company Galactic Suites announced its plans for a space hotel in 2007 with fanfare but no actual follow through to date [10]. The Russian company Orbital Technologies indicated plans for an in-orbit habitat in 2011, while a Japanese construction company indicated plans for building a lunar colony [11]. But to date such initiatives have been largely talk. Only Bigelow Aerospace has actually tested what seems to be viable technology in space to date.

Commercial Hypersonic Transportation

The skeptics of space tourism or space adventures have discounted the basic business plan of those seeking to simply fly people on suborbital flights so that they can experience weightlessness and see the Big Blue Marble against a black sky black drop. Those that discount this type of space tourism business plan suggest that there are a limited number of people who are willing to pay big bucks for such an experience and that once accomplished the repeat market would be quite small. The entrepreneurs who have looked into the future of spaceplanes or hypersonic planes have looked beyond space tourism. These Space 2.0 entrepreneurs have consistently suggested that the truly sustainable future business would not be flights to nowhere – i.e., suborbital parabolic flights that take off and land at the same point. They see the true potential as being hypersonic transport that could connect London to Sydney or New York to Tokyo in only a few hours.

These hypersonic flights for business tycoons and wealthy jetsetters would take off from airports and fly up to 50,000 ft at subsonic speeds. Then they would use rocket propulsion to reach speeds of Mach 3 to Mach 6 to reach an apogee of perhaps 80 kilometers and then descend and slow to subsonic speeds before landing half way around the world. New technology is being developed by NASA and others, such as an extendible needlelike nose system that would extend from the front of the spaceplane to avoid the generation of a huge sonic boom in the landing process. This type of extension system from the spaceplane's nose would create a thousand lower-level 'micro-booms' that would replace a thunderous clap of noise prior to landing.

Currently there is a dual path of development. One type of development is for supersonic aircraft that might fly at speeds such as Mach 2 or Mach 3. Then there are various spaceplane models that might fly at velocities on the order of Mach 5 or Mach 6.

There are quite a few questions that must be answered before true hypersonic services can seriously be started. Prime among these are the following:

- Is there a solid business case for hypersonic flights for truly long-distance flights of over 12,000 kilometers (7500 miles)?

- Are the considerations of speed, safety and cost sufficient to drive the development of spaceplanes (i.e., with speeds up to Mach 6) over the development of more conventional supersonic jets (i.e., with speeds of only Mach 2 or 3)?
- Which of the various technologies that are now available are the best ones going forward? The Japanese have made serious progress with H_2–0_2 hypersonic transport systems. Reaction Engines ramjet technology uses air from the atmosphere and makes it more efficient in that oxidizers do not have to be used in suborbital flights. Some of the spaceplanes to be used by Virgin Galactic use metallic fuels that spew particulates into the stratosphere. Performance, safety, reliability, cost and environmental impact are currently at odds with one another.
- Are there serious environmental concerns that apply to any of these types of vehicles in terms of their longer term operation in the stratosphere? Is the likely level of air pollution and particulates that would occur from flights through the upper reaches of the vulnerable stratosphere entirely too large if operated over multiple years? Would such hypersonic flights with apogees in the 80- to 100-kilometer range turn out to be too destructive to this fragile part of the atmosphere? Would systems such as Hyperloop or Maglev trains, especially if they are designed to use vacuum tunnels, turn out to be a better answer in terms of speed, safety and even longer-term cost?
- Can sufficient space traffic management systems be created to allow the safe operation of hypersonic craft through the protozone region of the stratosphere? What types of technology need to be developed and implemented in order to operate such hypersonic transportation systems in a safe and reliable manner?

Figure 12.6 shows the Skylon single-stage-to-orbit scramjet vehicle by Reaction Engines, the Japanese Hypersonic vehicle and SpaceShipTwo. Each of these vehicles have advantages and disadvantages that could advance the idea of hypersonic travel or even more cost effective travel to orbit. The future of hypersonic and space travel will be facing key challenges in the years ahead, and we are still lacking a

Fig. 12.6 *From left to right:* The Skylon Scramjet, Japan's Hypersonic, and SpaceShipTwo.

clear set of criteria for judging what is the best way forward.

Conclusions

There has been enormous progress made in the development of spaceplane and hypersonic transportation systems in the past decade. The 2020s will be a critical time for deciding what is the best path forward for the development of new transport systems for air and space. Never before have there been quite so many options that might be used to launch spacecraft into orbit, perform suborbital flights, and provide new options for hypersonic flight.

In 2004, when Burt Rutan, Paul Allen and Mike Melvill teamed up to win the $10 million Ansari XPRIZE, the next step forward into the world of space tourism and forthcoming public rides into space seemed clear cut and inviting. The prospect of hundreds of new citizen-astronauts was seemingly on the doorstep of that new tomorrow. The pathway was more difficult than it was thought at that time. A fatal Halloween accident in 2014 proved a serious step backward in the development of SpaceShipTwo. Sir Richard Branson, who had said that the spaceplane suborbital flights would carry him and his family on the first commercial flights, wisely deferred the development until tests proved his new vehicles to be truly safe. The development of commercial flights for citizen-astronauts since 2004 has been a story of one step forward, two steps back, and then one step forward again.

This chapter, however, tries to put developments since 2001 in perspective – going back to when the XPRIZE was established. In the last two decades there has been a broader rise in commercial space, or Space 2.0. We have seen the cross-fertilization of the aerospace industry and cyber-industry in new and innovative ways. This innovation has seen the rise of small satellites, new launch systems, large-scale satellite constellations, new applications in the protozone, on-orbit servicing, new manufacturing techniques and key new technologies in Earth station systems. The enthusiasm for space tourism and the media attention that has been heaped on celebrities going into space has helped the overall broad revolution in the space industry. This is not to dismiss the spaceplane initiatives to give rides to the stratosphere for high-rolling celebs or efforts to create private space habitats, but it is to say that the changes across the broad spectrum of the space industry must be considered as a whole.

References

1. Haines, L.: SpaceShipOne Claims X-Prize: Triumph for Private Space Vehicle. The Register, London (2004). https://www.theregister.co.uk/2004/10/04/spaceshipone_claims_x_prize/
2. Foust, J.: Second SpaceShipTwo Performs First Powered Test Flight. Space News, Alexandria (2018). https://spacenews.com/second-spaceshiptwo-performs-first-powered-test-flight/
3. Weitering, H.: Virgin Galactic's VSS unity space plane aces test flight, reaching mesosphere for the 1st time. Space.com (2018). https://www.space.com/41295-virgin-galactic-reaches-mesosphere-epic-test.html?utm_source=sdc-newsletter&utm_medium=email&utm_campaign=20180728-sdc
4. Pelton, J.N., Marshall, P.: Launching into Commercial Space. AIAA, Reston (2016)
5. The ESA-initiated future European space transportation investigations program. Accessed 29 July 2018. https://www.

researchgate.net/publication/301195167_The_ESA-initiated_Future_European_Space_Transportation_Investigations_Program_1994-1998
6. Merrill, J.: Alan Bond: the British engineer says he will revolutionise space travel – and he's building the rocket to do it. (2014). https://www.independent.co.uk/news/people/alan-bond-the-british-engineer-says-he-will-revolutionise-space-travel-and-hes-building-the-rocket-9587165.html
7. Op Cit, Pelton and Marshall, Chapter 6
8. Inflatable BEAM structure. NASA. Accessed 30 July 2018. https://www.nasa.gov/mission_pages/station/research/experiments/BEAM1.jpg
9. Wall, M.: Private space habitat to launch in 2020 under commercial spaceflight deal. Space.com (2016)
10. Gordon, S.: Space hotel to open to tourists. Daily Mail (2009). http://www.dailymail.co.uk/travel/article-1224629/Galactic-Suite-Space-Resort-hotel-2012-First-space-hotel-open-tourists.html
11. Boyle, R.: Luxury getaways of the future: visit orbital technologies' space hotel. Pop. Sci. (2011). https://www.popsci.com/technology/article/2011-08/luxury-getaways-future-visit-orbital-technologies-space-hotel

Space 2.0 Economic, Business and Regulatory Issues

13

Introduction

Dr. John Logsdon, who directed the Space Policy Institute at George Washington University for many years, had a favorite saying when he lectured to the International Space University. This admonition was: "The regulators always win."

This pithy caution was Dr. Logsdon's way of saying that even those private space enterprises with the best technology and the best financial backing were not going to be successful if they were blocked from proceeding due to a lack of enabling legislation or governmental regulatory approvals. Government approvals can trump exciting technology or billions of dollars in financial banking. However, potential customers and market success is just as important as good technology.

Communications satellite companies, for instance, still need to obtain frequency allocations, get orbits authorized, and complete interference coordination under International Telecommunication Union (ITU) procedures. They also must get national landing licenses with all the countries where they intend to operate. All space application businesses depend on authorizations to launch satellites and assure launching nations that due diligence has been conducted against creating orbital space debris.

Such authorizations needed by commercial organizations to operate satellites in space must come from governments at the national and international level. All companies are subject to the provisions of Article VI of the Outer Space Treaty. This provision has the force of law in over 100 countries, and, in particular, it applies to all spacefaring nation states that have ratified this treaty. The provisions of Article VI state: "States Party to the Treaty shall bear international responsibility for national activities in outer space, including the Moon and other celestial bodies, whether such activities are carried on by governmental agencies or by non-governmental agencies…" [1].

This is to say that under the Outer Space Treaty and its four other subsidiary international agreements and conventions State approvals by the national governments are needed. Also commercial space entities must comply with the

J. N. Pelton, *Space 2.0*, Astronautical Engineering,
https://doi.org/10.1007/978-3-030-15281-9_13

guidelines of the ITU and other processes or guidelines developed within the U. N. Committee on the Peaceful Uses of Outer Space and ratified by the U. N. General Assembly.

In short, commercial space initiatives are still constrained by a number of treaties, laws and regulations. There are constraints that apply to accessing spectrum and getting approved spectrum and orbital locations that are licensed to them by governments. There are other areas of concern as well. These include such aspects as protecting against the hazards of existing orbital space debris, meeting due diligence to assure their licensing agency that they will not themselves create orbital space debris, provide assurances that their satellites will not create undue RF interference, and more.

Sometimes these constraints have the force of law and in other cases it is simply voluntarily following norms of behavior that have developed over time. The bottom line is that success for Space 2.0 ventures still requires more than innovative technology. Regulations are also a key part of the equation.

The NewSpace ventures that succeed must have a viable business plan, customers to purchase their service or product, as well as regulatory approvals, licenses, and authorizations.

Later we will address the critical importance to financial success that is tied to the requirement that global satellite communications companies' need for landing licenses in all of the countries in which they operate. Those that think that the key to success in operating a space business is simply a matter of having access to an exciting new technology need to think again. One needs access to technology to start. Then that company needs access to financial capital, a solid business plan and a number of regulatory approvals at the national and international level. Getting all of these component parts right and having them work in harmony are essential to making a viable space business. Many of these elements involve years to achieve (i.e., designing and building the space segment, designing and building the ground segment and user terminals, raising the capital, getting the license for the frequencies and orbital location, completing the ITU coordination procedures, getting the national landing licenses, recruiting the staff, marketing the services and more). It is possible to get many elements of a successful business right but still fail because one of the critical elements had not been completed successfully or was achieved too late.

This chapter reviews some of the key regulatory, business and financial constraints that apply to carrying out business in space. A regulatory turndown can be as daunting as a launch failure.

International Regulatory Challenges

Most space-based services and products involve the design, manufacture, launch and operation or deployment of a spacecraft or a high altitude platform system. Such a launch requires approval at the international level through the various processes of the ITU to obtain appropriate frequencies and completion of spectrum and orbital coordination with regard to potential frequency interference to other terrestrial or space facilities.

This process starts with the proposed registration of the satellite and its orbital parameters through a national administration that is a member of the ITU. The ITU then sends out these technical parameters to all of its members to seek input as to whether there are concerns with regard to potential interference. Its staff considers if the proposed registration of the frequencies to be used is consistent with the global allocations approved by the World Radio Conference.

In the early days of space applications this process was straightforward and relatively easy to conclude. Today the process is much more complicated. Satellites in geosynchronous orbits have protected status with regard to non-GEO satellites. Further, there are now proposals for some LEO constellations with several thousand satellites and thus the technical consideration of intersystem coordination is far more difficult.

The proposal to establish a satellite system typically begins with the effort at the national level to get approval for the licensing of the proposed satellite network. The national licensing authorization process may take many months or even years. Such delay at the national licensing level has led to the filing of so-called 'paper satellites' with the ITU, to claim certain locations and spectrum ahead of such pending filings in other countries.

Such competing accelerated filings with the ITU were seen as a sort of 'flag of convenience' initiative. The purpose was to stake a claim to frequencies and orbital locations ahead of filings by others who might apply in the United States, Europe, Japan or other countries that have a more deliberate and slower approval process.

Countries such as the Kingdom of Tonga, with its so-called Friendly Skies Corporation, Gibraltar and Papua New Guinea were seen as making filings for paper satellites. In response to these filings and complaints with regard to flags of convenience filings, changes were made to the ITU coordination processes to discourage these practices. The ITU fees for such filings were increased. New requirements were instigated that required due diligence filings to show that certain financial resources were available to build and deploy the satellite systems in question. Assurances had to be provided that contracts had been awarded to satellite manufacturers. Proof of other due diligence steps that were being made to actually build and deploy the system in question were also required. These steps have discouraged the filing of paper satellites, but these processes have also made it more difficult for new start-up ventures such as Spire, Skybox, Planet Labs, OneWeb and other new providers of satellite networks to complete the required ITU processes.

National Regulatory Approvals Around the World

The allocation of new frequency bands for space application services and the development of new technology can lead to a rush to file applications to use new bands of spectrum. In the 1990s there was new interest in the possible use of the Ka-band spectrum to provide satellite communications services. The NASA ACTS satellite had demonstrated the feasibility of using this band for communications services. The ACTS satellite demonstrated the technical capability of Ka-band satellites to cope with rain attenuation, especially in the

28 GHz band. The U. S. FCC indicated that it would review applications for satellite systems to use the 28 GHz uplink and the 18 GHz downlink bands. Thus there was a glut of applications for Ka-band satellite communications services.

Of those original applications to the FCC in the United States only the Wild Blue Ka-band system was ultimately deployed. Many of these proposed systems were abandoned, especially after the planned Teledesic satellite network declared bankruptcy. This history is perhaps relevant as one looks at the companies hoping to build and deploy large scale constellations of satellites to operate in the Ku, Ka and Q/V bands to provide new satellite links for Internet services to underserved areas of the world.

The very strict requirements of some governmental licensing authorities, such as the FCC in the United States, have ended up making the cost of filing an application for the licensing of a new satellite system several million dollars. The high cost of these applications and the long period for review and approval has tended to discourage the filing of new satellite systems.

Recently some governments have sought to address this issue by finding ways to shorten approval processes and to make filings less onerous. Some have even sought to create financial incentives to attract Space 2.0 companies to their country. Luxembourg has created legislation that would allow space mining companies to retain resources obtained from outer space. They have also created a new space industry initiative backed by a $200 million capital investment fund to attract properly vetted space industries to relocate to Luxembourg. This is on top of the attraction of not having corporate income taxes. New Zealand has created a special program to attract space entrepreneurs to come to their country. Recently the United States has prepared a series of initiatives aimed at keeping Space 2.0 industries in the United States. Secretary Wilbur Ross, at a Hudson Institute forum on space policy, outlined how applications related to new remote-sensing projects are now given approval within 90 days and how the new Department of Commerce responsibilities in the space business sector will be aimed at attracting or keeping Space 2.0 industries in the United States.

Despite these initiatives to attract new space industries to some countries, there is another aspect of the licensing process that can serve as barrier to new space-based services. Most countries have established laws that require licenses to install land cables or operate satellite systems within their territory. The World Trade Organization has created a process that seeks to open up telecommunications to more competition and to ease barriers to providing competitive telecommunications and information technology services around the world among its membership.

Nevertheless, most countries require some form of revenue sharing when a company initiates services within its borders. The question is how these new satellite constellations will operate and be licensed locally? How they will be required to share their revenues under the requirements set by local regulatory agencies that oversee telecommunications and IT service entities going forward?

There are today over a dozen new constellations that are planned to be built, deployed and begin to offer services, and most of them anticipate that they will operate in scores of different countries. In many instances there are special and unique requirements set by each country, and those must be met in order to operate within each country. Some countries require a local office and a certain level of expertise and staff for that office. Other countries require that connection to a satellite network must be through a local telecommunications or networking office and that a certain amount of the revenues must be derived from local companies. The point is that although there are general guidelines that the World Trade Organization has set to encourage the possibility of international competition, there is no single set of requirements or uniform standards that apply.

The governments of some countries are far more protective of their national telecommunications and networking companies than others. The countries with the lowest levels of economic development often have the most restrictive requirements to protect their local telecom and IT industry and are the most fearful of outside competitors who seek to offer new services locally. On top of these concerns, governments in developing parts of the world also have the highest tariffs, to limit the import of such items as computers, communications terminals and other forms of electronic equipment.

There are organizations that have been formed to champion the cause of low-cost communications and the use of new satellite or computer technology to provide low cost education, training and health care systems in rural and remote parts of the world. One organization in particular has developed a white paper that explains the economic, education, training, medical and health care advances that can come with new low cost satellite systems technology. This white paper, which was developed by the "Geeks without Frontiers," has explained why rural satellite services are important to developing countries and how the new low Earth orbit satellite systems can provide new types of benefits to countries that have limited rural connectivity. It is important to note that governments that have long-standing policies to defend against outside exploitation may see such innovation and new types of electronic services in a much different light [2].

The bankruptcies of the Iridium, Globalstar and ICO companies that sought to provide mobile satellite services in the 1990s was tied in significant part to problems of regulatory authorization to operate in all the countries where it intended to operate. (See Fig. 13.1.)

The unexpected costs of sharing revenues and the difficulty of negotiating landing agreements with many scores of countries was certainly a key factor. Negotiation of these landing licenses likely had as much as to do with the financial failure of Iridium and Globalstar as it had do with the technical design of these networks [3].

There do seem to be serious questions as to whether all of the new satellite constellations systems that are now contemplated can be economically viable. The key question is whether the regulatory models and revenue sharing issues have been adequately addressed in all of the new systems that are now planned. The O3b satellite system, which sought to provide new communications and IT services to the equatorial

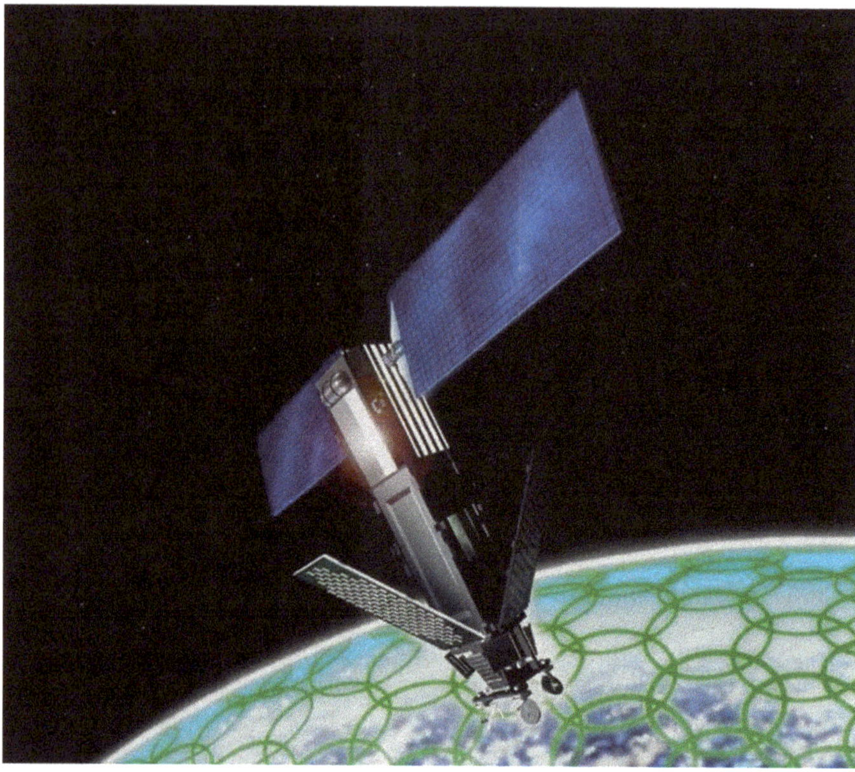

Fig. 13.1 The Iridium satellite design for the Global Mobile Satellite System. (Graphic courtesy of the Iridium Satellite System.)

regions of the world, has successfully found sufficient revenues and gotten regulatory approvals to make this MEO constellation satellite system work. But the creation of one new system by an established satellite operator such as SES of Luxembourg and making it work is much different than starting over a dozen new satellite systems and each of them successfully sorting through the regulatory complexity that is always informed on a country by country basis.

The bottom line is that getting regulatory approvals remain a very key part of the success of all of these Space 2.0 ventures.

Flag of Convenience Arrangements and Short-Circuiting Regulatory Filing Requirements

One shortcut that has been tried to avoid many of the delays and difficulties associated with getting a license for a new satellite system has been to get a flag of convenience-type license from a country that agrees to help in this regard. The country of Panama has for years served to license ships at minimum cost and easy regulatory process. There are national entities that have undertaken this sort of flag of convenience role with regard to the filing of satellite systems. The Friendly Skies Corporation, known as Tongasat, made an agreement with the King of Tonga to file various so-called paper satellites with the ITU for a variety of purposes. The nation of Papua New Guinea filed on behalf of another satellite system on the understanding that they would get access to a satellite transponder that would give them health and educational services. Mansat, on the Isle of Man, and other entities have also served to provide regulatory or taxation advantages to those that intend to develop new satellite systems.

This type of short-circuiting of detailed technical, financial and regulatory review can lead to problems later on. The ITU, under pressure from its member states, has taken steps to avoid the use of such processes to proceed with filings of satellite systems that have not had a thorough technical, financial and regulatory vetting. These new processes require that filings be backed up with additional financial, technical and contractual information documentation that go beyond the original filings allowed a few years ago. Further, the ITU fees for all filings now come at higher cost. These steps have lessened the number of paper satellites, or filings for satellites that were never actually intended to be launched. This also means that all satellite filings are more carefully considered by national administrations from a technical, regulatory and financial perspective before they are filed internationally.

Capital Financing

The regulatory requirements with establishing a satellite system, especially if it involves global service, can clearly be daunting. But because of the long lead times to design, build, launch and begin operations the financial risks can be quite large. These conditions can become even more exaggerated when the project is trying to establish a revenue stream in a new market such as Internet services in rural and developing regions of the world and the new satellite system involves the manufacturing and launching of hundreds of satellites before the revenue stream can begin to pay off large amounts of capital debt.

One large-scale satellite constellation project started as the Calling Communications project in the late 1990s. It was formally incorporated as the Teledesic in 2000 and ended its existence in 2002. The author at the time this project was proposed noted the problem of building and launching quite so many satellites in an untested Ka-band frequency with a huge capital expenditure many years before a significant revenue stream could be realized.

Investors in the ambitious project to build an Internet in the Sky included Microsoft CEO Bill Gates, cell phone magnate Craig McCaw, Boeing, Saudi Prince Alwaleed bin Talal and others. They managed to raise over $200 million in financing, but the design for the network was in constant flux from the time the project was first conceived by designers from the University of Colorado and a venture capitalist from California in the late 1990s. The initial plan was first conceived as an LEO constellation that would consist of 840 Ka-band satellites plus 80 spares. Ultimately the project was canceled in October 2002 after millions had been spent in designing and redesigning the project. A contract with Thales Alenia to build 30 of the satellites

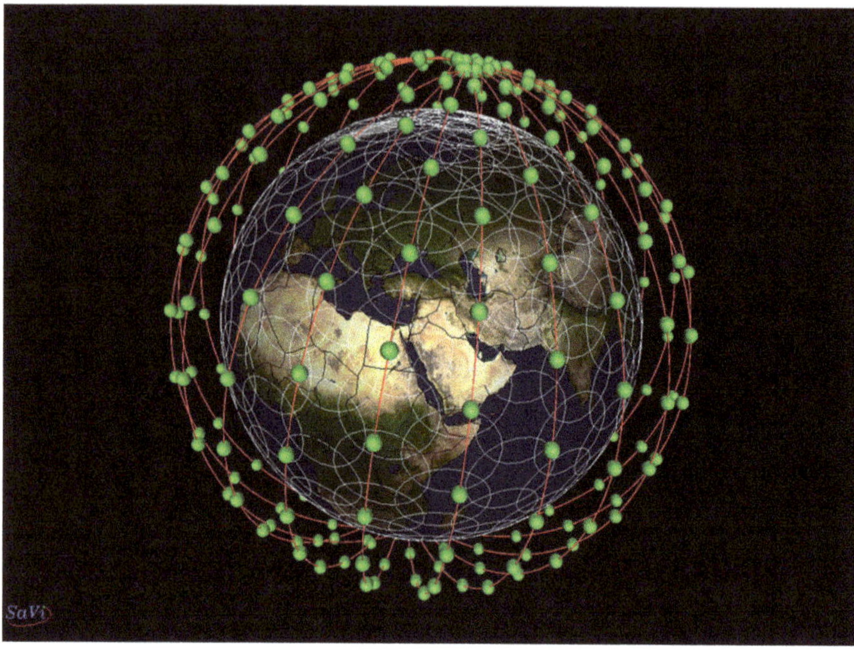

Fig. 13.2 Ka-band satellite constellation design. This 920 LEO satellite constellation was never built due to projected revenue stream difficulties. (Graphic courtesy of Teledesic.)

was abruptly canceled. In announcing the cancelation, the press release stated that due to an "unprecedented confluence of events in the telecommunications industry and financial markets….it had decided to suspend work because the returns to shareholders weren't commensurate with the risk [4]."

Changing Satellite Business and Economic Models

The tendency might be to think of all types of satellite business markets as much the same. The truth of the matter is that each type of satellite market tends to be quite different. Domestic satellite communications services depend on the strength of the domestic economies. Further the market for television distribution, direct broadcast satellite television and fixed satellite services also differ in terms of the design of the satellite, the frequency bands they use and the type of consumer or user terminals that are used. In the case of regional or global satellite services there are dozens of submarkets, and each of these tends to be different. There are many market divisions, such as direct broadcast satellite services, television distribution services, direct audio broadcast satellite services, fixed satellite services for telephone, data distribution, enterprise network connection, business television and backhaul connectivity for mobile services. Mobile satellite services also have subdivisions. Further there are yet other markets to respond to military or defense-related services that might be contracted to provide services on a dual use basis. And this represents only the market divisions related to telecommunications, networking and broadcast services.

There are also different types of markets and technologies for those that are seeking to provide remote sensing, Earth observation and surveillance services. Some of the key differences in the remote-sensing area involve the technical capabilities related to radar, infrared, optical and ultraviolet and whether the data is hyperspectral or not. Precision navigation and timing services are increasingly branching out, based on new software-based applications as more and more types of markets are served.

In short, the world of satellite services is changing rapidly as new markets are being defined. As new ways are being developed to design and build satellites at different cost levels, the same is true with regard to the design, manufacture and cost of consumer transceivers and terminals, where the nature of markets has changed dramatically.

On top of all these dramatic changes to satellite technology, ground station technology, satellite services and new types of market demand, there are also huge shifts in the competitive technology. Dramatic changes in the cost, performance and availability of fiber optic networks, broadband cellular systems on the ground and even new systems such as high altitude platform systems and UAV capabilities can open up the possibility of alternative systems and services on the ground, in the stratosphere and even between LEO, MEO and GEO satellite networks. Market conditions and technical options are in a state of flux as never before. Because of the lead times associated with the design, building, licensing and authorization, deployment and testing of satellite systems, there is greater volatility in satellite markets than ever before.

Some of the key drivers of change in the world of satellite services, associated markets and lower cost systems that have arisen in the past few years include the following:

- Creation of new and more efficient ways to design, manufacture and test satellites, including such aspects as additive manufacturing, miniaturization of key components and use of off-the-shelf components.
- Development of new designs and technical performance capabilities for ground-based user terminals. This is especially true with regard to the new use of meta-materials to create lower cost flat antennas with the ability to provide electronic beams and thus electronic tracking of low Earth orbit satellites in constellations. This is a more cost efficient way of operation rather than implementing antenna systems that require physical tracking of non-GEO satellites.
- Advancement of a wide range of innovations in the launch industry to make the launch of satellites more reliable and much more cost effective via such improvements as reusable launch vehicles, vertical integration of supply changes, new types of smaller launch vehicles from start-up launch ventures and other innovations such as carrier systems to allow launch from higher altitudes (i.e., the Eve and Stratolaunch carrier planes that allow boosts in the stratosphere).
- Creation of new forms of market competition among high throughput GEO satellites, smaller satellites in large-scale constellations in low Earth orbit and creative market

responses by conventional satellites with lower throughput rates.

- Implementation of major advances in efficient encoding and forward error correction systems that will allow satellites to operate up to 10 bits per hertz efficiencies in the relatively near future.
- Improvement in sensors that allow miniaturized devices on small satellites to provide hyper-spectral sensing capabilities.
- Advances in techniques to cope with rain fade issues that will allow the effective use of the Q/V spectrum bands (i.e., 48 GHz/38 GHz) to provide telecommunications and networking services via satellite.
- Flexible new ways of raising capital for new satellite ventures using the Internet and unconventional financing methods.
- New, more flexible and more user-friendly methods of governmental regulations and licensing procedures that countries seeking to promote Space 2.0-type ventures are now using to encourage new filings for space-based services.

Recap of Key Regulatory Changes

The main thrust of regulatory changes in the past few years have been to accomplish the following goals:

- Regulatory action, especially within the World Trade Organization (WTO), to encourage more competition with regard to space-based services and to discourage nations from creating tariff or non-tariff barriers to entry into these markets by creating undue restrictions.
- Reduce the number of paper satellites and seemingly unfair practices that might occur from flags of convenience filings for new satellite systems that cut corners with regard to technical, financial, contractual, or regulatory provisions. Some of these reforms involve ITU fees and practices and others involve changes at the national governmental level.
- On the other side of the spectrum, new reforms and less demanding legislation has been enacted in some spacefaring nations and nations wishing to encourage Space 2.0 initiatives. These actions have been undertaken so as to not be overly restrictive in the licensing process with regard to how satellite systems are designed, manufactured and deployed. There have also been legislative and regulatory actions to possibly encourage new space ventures such as solar power satellites, on-orbit servicing and space mining.
- Efforts to restrict the creation of new space debris and create more stringent due diligence procedures against the creation of space debris by prelaunch procedures and to institute fines or penalties if the guideline of removing spacecraft from orbit within 25 years of end of life is breached.
- Finally there is the issue of orbital space debris and space traffic management. This includes the issue of some oversight with regard to subspace, the so-called protozone. Most of the concerns in this regard are directed toward increased safety, prevention of collisions and avoidance of new space debris being created. It should be recognized, however, that there is a matter of efficiency of operations of aircraft

flight that is also involved. The current procedure involves shutting down of air corridors as launch operations or special ascents occur. It is argued that if there were an integrated air and space traffic control system, diversion of flights and aircraft diversions could be minimized and overall integrated operations made more efficient [5].

Despite what might be called progress in the above areas, there remain many issues of safety and debris minimization and active debris removal that remain to be addressed and resolved. Further there is dispute concerning different interpretations of the Outer Space Treaty, the Moon Agreement and the Liability Convention in areas related to active debris removal, the guidelines and practices related to on-orbit servicing and space mining activities. The hope is that between the good offices of the InterAgency space Debris Committee (IADC), the COPUOS and its Working Group on the Long Term Sustainability of Outer Space Activities (LTSOSA), and other consultative bodies that progress can be made in these areas.

Conclusions

The tendency of technologists is to assume that the most important trends and developments in space systems come from new invention and technical breakthroughs. Regulatory officials and legislators tend to the view that their actions guide and largely define the future. Bankers, investment officers and business executives tend to believe that their decisions are the most important. The truth is that it is a combination of technology and operational process, regulatory and administrative rules, laws, and treaties, plus business and investment decisions that determine the future of space systems. Since space activities by their very nature are international in scope, they are complex. The world of space is not only international but interdisciplinary, and anyone in this field needs to know something about technology and technical standards, about applicable law and regulations, and about business practices and financial investments.

This book has largely addressed space systems and services. It has examined how the world of Space 2.0 is redefining the nature of space businesses, who the major players are today and how they may indeed change tomorrow. This chapter, however, has sought to provide some insights into the importance of regulatory processes and business and financial processes, which are critical to understanding the space industry and why it is changing so drastically in the last few years in the face of Space 2.0 initiatives. It is the new way of thinking that has come out of Silicon Valley and the cyber-industry that has brought a new dynamism and an infusion of new talent and ideas to the space industry. The space industry is indeed being reinvented by new technology, new regulation, and new patterns of investment.

References

1. 2222 (XXI) Treaty on Principles Governing the Activities of States in the Exploration and Use of Outer Space, including the Moon and other Celestial Bodies, adopted December 19, 1966. http://www.unoosa.org/pdf/gares/ARES_21_2222E.pdf

2. Connectivity is the revolution. http://geek-swf.org/initiatives/. Accessed 8 Aug 2018
3. Mellow, C.: The rise and fall and rise of Iridium. Air and Space Magazine, 4 Sept 2004. https://www.airspacemag.com/space/the-rise-and-fall-and-rise-of-iridium-5615034/#0i0eXz80cUhwdMpz.99
4. Bloomberg News.: Teledesic suspends work on satellites. Seattle Times, 1 Oct 2002. http://community.seattletimes.nwsource.com/archive/?date=20021001&slug=teledesic01
5. Jakhu, R., Pelton, J. (eds.): Space traffic management and coordinated controls for near-space. In: Global Space Governance: An International Study, chap. 13.4.2, pp. 318–319. Springer, Cham (2017)

The Way Forward 14

The Blossoming of Space 2.0 Enterprises

The world of NewSpace or Space 2.0 enterprises has blossomed in just a few short years. Commercial space activities now dwarf governmental space programs. The scope of these activities just keeps on expanding. Innovative new space businesses keep increasing in original and sometimes even unsuspected ways. When the cyber-industries found their way into the world of aerospace it encountered a fertile ground in which to plant new ideas, grow new enterprises and reform and revitalize institutional and regulatory processes.

The chart in Fig. 14.1 shows just how sophisticated and diverse commercial space-related activities are today. Currently private space businesses outstrip public space activities – both civil government and defense – and by a large margin. This ratio of governmental to commercial ratio for 2016 is shown in the chart. For 2016, governmental space activities, including defense space activities, totaled $83 billion (U. S.) versus $261.5 billion (U. S.) if commercial products and services are considered as a whole. This adds to a global total of $344 billion (U.S.) [1].

The rapid advent of new launch and new satellite applications, and other Space 2.0 activities causes the ratio to tilt even more to the commercial world. The merging of space-linked activities with cyber and Internet-based activities will fuel the growth of both of these industries. Internet of Thing-type services that are linked to satellite connectivity is but one of the ways in which this phenomena is true.

This significant shift to commercial space activities is driven by many factors. These include new technological innovation and new ways of approaching the management of these space systems. The biggest change is the reshaping of entrepreneurial thinking about commercial space opportunities and the regulatory and business reform that has followed.

The latest comprehensive information for 2017 now available indeed shows this continuing shift in favor of commercial space activities with reduced governmental space expenditures. Thus the global space economy

J. N. Pelton, *Space 2.0*, Astronautical Engineering,
https://doi.org/10.1007/978-3-030-15281-9_14

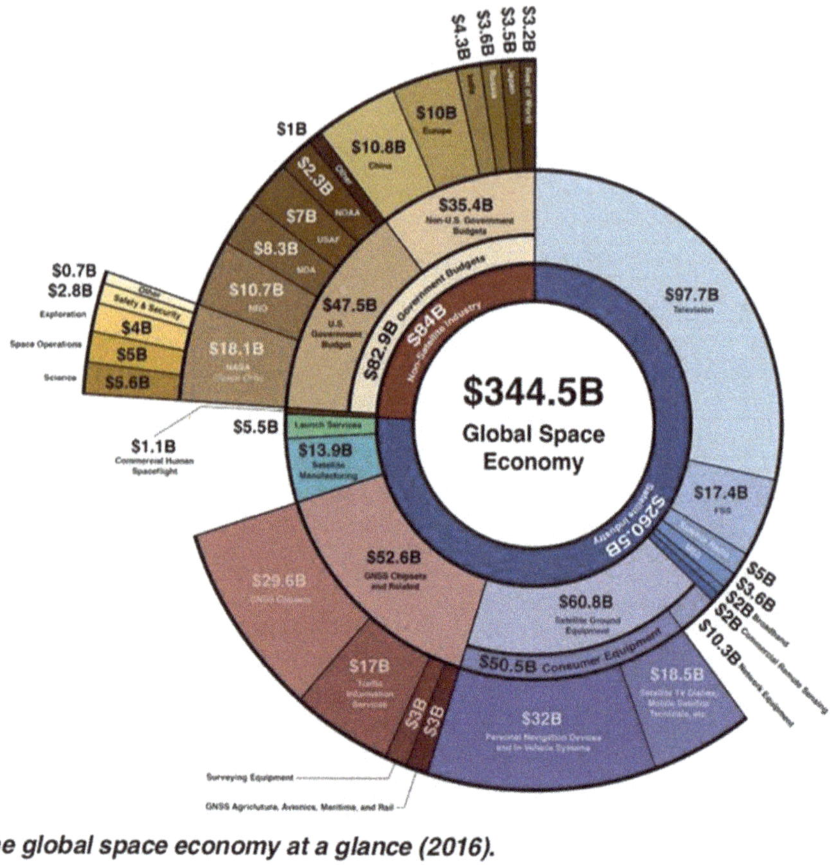

Fig. 14.1 This chart shows the shift from governmental involvement in space to commercial involvement in space. (Graphic courtesy of the Satellite Information Association.)

for 2017 totaled $348 billion. Of this amount $79.3 billion (U. S.) related to governmental and military expenditures while the remainder related to all combined commercial activities at $268.7 billion (U. S.) Thus total governmental expenditures were down about $4 billion (U. S.) and total commercial space activities went up by about $8 billion [2].

These changes in part are fomented by entry into commercial space business of new players. New regulatory regimes, standards, services and even technologies have served to lower the barriers to engaging in space business enterprise. The more that Space 2.0 began to be like the world of computer services and the social media enterprises, the more the door was opened to new types of space services. These changes also lowered the amount of capital financing and shrunk the size of the companies and their expert staffing needed to compete in this new world. Satellite communications services, for instance, are today at least ten times more efficient than they were a decade ago. The bulk of these gains have not come from better hardware. No, the largest improvements in throughput of bits per Hertz have come from better software, improved coding techniques and better forward error correction systems.

The explosion of activities that now populate Space 2.0-type ventures is almost mind boggling. Let us explore the new dimension of commercial space activities.

The Ever Widening Range of Space 2.0 Enterprises

NewSpace activities are now envisioned in three different areas. These include the so-called protozone region (20 to 160 kilometers in altitude), near-Earth space and some of the most ambitious new space enterprises, which envision deriving benefits from the far reaches of outer space. The range of activities includes at least the following:

- large-scale small satellite constellations in non-geostationary orbit.
- high throughput satellites for telecommunications, broadcasting and IT services.
- hyper-spectral cube sats and other types of small sats for remote sensing.
- hypersonic transport that would employ spaceplanes to fly through the protozone.
- The space tourism and space adventures business.
- new low cost and reusable commercial launcher systems. (This includes systems such as Stratolauncher by Vulcan Industries, carrier planes by Virgin Galactic, reusable launch systems and small satellite launcher systems, including those that add a launcher stage to spaceplanes.)
- other advanced space transportation systems such as those using large-scale space tethers, solar sails, nuclear propulsion, advanced electrical ion systems and other new approaches.
- high altitude platform systems (for communications, broadcasting, remote sensing, law enforcement, fire and crime detection, etc.).
- on-orbit space servicing and retrofit of satellites, including active space debris removal, redeployment of satellites that did not achieve proper orbit, etc.
- dark sky systems for research, deployment of small satellites to orbit using electric ion propulsion, etc.
- robotic freighters operating in the protozone for global air freight activities.
- space processing and manufacturing and reclaiming of resources from outer space (i.e., space mining).
- large-scale constructions in space including the building of planetary defense systems against cosmic hazards (solar-shields in L-1 against coronal mass ejections, space systems to combat asteroids and comets, etc.).
- private space habitats and private orbiting research facilities.
- solar power satellites that can beam back clean energy from space 24 hours a day.

These are only some of the systems where commercial ventures or serious research and development are underway. Arthur C. Clarke, in discussing forecasts of the future, had noted that most things we can conceive of we may one day be able to accomplish. His three laws of prediction, described following, are useful reminders that we should not be too quick to discount what might be accomplished by new technology in the future [3].

- *First Law:* "When a distinguished but elderly scientist states that something is possible, he is almost certainly right. When he states that something is impossible, he is very probably wrong."
- *Second Law:* "The only way of discovering the limits of the possible is to venture a little way past them into the impossible."
- *Third Law:* "Any sufficiently advanced technology is indistinguishable from magic."

If one considers where we have come in the past hundred years it is truly amazing. We have seen the development of airplanes, computers, rocket launchers, nuclear fission and fusion, antibiotics, stem cell research, and even the invention of spandex. It seems that we have just begun to exploit a tiny fraction of human potential if we do not end up making Earth uninhabitable or annihilating our species. Space 2.0 seems to be a giant part of that human potential.

In many ways the future of Space 2.0 seems likely to be shaped by just a few things. These are, in essence: technology, regulatory policy and governance, and economic, social and cultural potential.

The Potential of Technology

For years space applications have made advances in such areas as satellite communications and broadcasting, remote sensing, space navigation, weather satellites and more. Progress has been made on all fronts. Launch vehicles have become more capable of launching larger payloads to space. Antennas have become larger, achieved higher gain and been pointed more accurately to achieve higher throughput. Sensors have become smaller while also achieving higher spatial resolution. Ground systems have also become more capable and cost efficient.

In the area of satellite communications important gains have been achieved in all three of the vital areas available for performance advancement. Thus these satellites today have much greater power, access to much wider bands of spectrum through use of new bandwidth provided by higher radio frequencies and digital processing complexity that allows much greater throughput of bits per Hertz.

Comparable advances have been made in remote sensing and meteorological satellites through the use of more and more capable sensors that have shrunk in size while improving the resolution of space-based imaging. Precise navigational and timing satellites have also advanced through the use of more accurate atomic clocks. In all areas of satellite operations, in space and on the ground, improved software and digital processing has advanced performance, accuracy and cost-efficiency.

There is nothing to suggest that the improvements in satellite hardware and software have run their course. In short, the ability to design and create better hardware, exploit higher bandwidth and create improved software, achieve higher performance and allow improved cost-efficiency all seem very much likely to continue for some time to come.

This is not to suggest that there are no limits to growth. Today, perhaps the greatest barriers to future growth and improved service come from two sources. These are the limits that come from orbital space debris and access to frequency spectrum. This is, of course, to put aside the political and military dimension in that warfare and hostilities in space could set back progress in space in almost unimaginable ways.

Significant Increase in Space Debris

The current situation is that the number of operational satellites in orbit might rise from 1,500 spacecraft to over 15,000. The amount of debris over 10 centimeters in diameter currently being tracked is over 22,000, and projections of new significant orbital collisions are that these will occur perhaps as frequently as once every five years, with this likely to give rise to over 2,000 new debris elements. This future profile of space debris has led to recommendations that there be new active space debris programs to remove the largest debris elements, such as Envirsat, from low Earth orbit as quickly as possible with at least 10 such removals a year. There is serious concern that the so-called Kessler syndrome can unfold in coming days and endanger all forms of space deployments in the future. There is grave danger that inaction to reduce space debris would ultimately lead to runaway buildup of debris and a lethal avalanche of space junk unless action is taken. Currently harmful space junk thus seems to pose the greatest risk to low Earth orbit operations.

Satellite RF Spectrum Concerns

The other major significant technical concerns for the future relate to allocation of RF spectrum to meet future satellite application needs. Currently there is enormous demand for spectrum to satisfy the needs for terrestrial broadband and the streaming needs that will come with 5G and 6G broadband systems that are likely to claim frequencies that have been reserved for satellite services. There are many ways to make satellites' use of frequencies more efficient and achieving even more bits of throughput per Hertz, but the loss of key spectrum allocation could hamper the development of satellite services in the future. Further the demands for spectrum to support high altitude platforms systems and UAV-based services could also make inroads into satellite spectrum needs.

The Reinvention of Space Processes

Space 2.0 initiatives have grown out of technological innovation and new ways to design, deploy and use space systems. Economic efficiencies, miniaturization of design and new ways of delivering space-based services have, in many ways, come from the cyber-industries. "Silicon Valley meets and transforms the aerospace industry" is the big story of the last few years of the global space industry.

The story can in part be told in the names of the space billionaires who are intent on reinventing an industry to be more nimble and innovative, an industry that was for decades dominated by the supply of spacecraft and launchers to governments, military forces and large and well-established aerospace organizations. Today the key names in this industry are different. They are now people like Elon Musk of SpaceX, Jeff Bezos of Blue Origin, Robert Bigelow of Bigelow Aerospace, Sir Richard Branson of Virgin Galactic, Bill Gates involvement in Kymeta, and Paul Allen's Vulcan Industries and Stratolaunch.

Yet the realization of these innovations via NewSpace infrastructure and new kinds of startup companies requires more than just new ways of thinking and new leadership. In many cases there also had to be other innovations to make these ventures possible. Thus there had to be changes to regulatory frameworks, new types of licensing processes and new ways for startups to get their financing such as "Kickstarter" and other ways to turn an idea into an actual business.

Despite the many new ways of designing and building satellites, new kinds of exciting Earth station technologies, new launch vehicle designs and new markets to be served in the omnipresent world of the Internet, much more is still to be done to make digital access via satellite even more pervasive.

There still remain many tariff and non-tariff barriers to entry into domestic markets around the world and especially in rural and remote areas. This is not only a matter of overcoming tariff barriers, incompatible technical standards, landing-license constraints or other regulatory barriers and obstacles. In many areas of the world there are still fundamental barriers to getting what seem to most as a fundamental right to service. There are still billions of people who cannot obtain remote telecommunications, broadcasting, networking, weather services, navigational, remote sensing, access to education and health services and basic governmental services. This is because even more basic needs cannot be met. In these areas of the world, there remains a lack of access to electrical energy or even more fundamental problems such as a lack of food, potable water and protection against virulent disease.

The Economic, Social and Cultural Potential of Humankind

The world still faces many challenges at the fundamental level. These include climate change and the ability of nations to provide their citizens with basic physical needs that include water, food, education and health care. Space services by themselves cannot be a panacea. The lack of electricity, lighting, food and water, education and health care, and a source of jobs and livelihood, must be seen as a holistic problem of which vital satellite services are only a part. The U. N.s seventeen Sustainable Development Goals for 2030 have actually attempted to create a global framework for meeting overall needs, but the road to success remains long and difficult to travel.

Satellite services can provide help in meeting many of the U. N. Sustainable Development Goals. The full and effective use of space services to meet these goals requires some recognition that satellite services are more than an add-on frill, but in many ways are vital to human survival in the 21st century. The central role that space services can play in meeting these goals is provided in the Chart 14.1.

Regulatory Reform by Governments to Encourage NewSpace Enterprises

The reinvention of the space industry has many dimensions. The most important dimensions include the development and application of NewSpace technologies, new launch systems and new ground systems and user facilities, new operating techniques and effective use of artificial intelligence and smart algorithms that allow for more effective and safer deployment of new technology, new spectrum bands, better systems to cope with space debris, new and innovative ways to finance NewSpace initiatives and better regulatory systems and space governance systems to allow for the allocation and use of radio frequency spectrum at the national, regional and global level.

As NewSpace industries are created and new types of spacecraft and launch systems are developed it will be crucial for regulatory systems to adjust to these new requirements and to find efficient and effective ways to license these services. In this way, they can be safe, well-coordinated so as to not create interference between the systems, and avoid collisions that would lead to increases in space debris that would be harmful to all types of space operations.

Concerns About Future Hostilities in Space

As noted in Chapter Nine all the progress that has come from Space 2.0 initiatives could be disrupted by hostilities in space or the launching of weapons of mass destruction into Earth orbit. Since the beginning of the Trump administration in the United States there have been many changes, such as the recreation of a National Space Council. There has been the issuance of at least three National Space Policy Directives as provided in the appendices to this book. Finally and perhaps most ominously there has been proposals and discussion about the creation of a U. S. Space Force.

Many space policy and law people around the world are concerned that these policy shifts as well as new military and strategic efforts in space on the part of China, Russia and the United States could disrupt the peaceful uses of outer space. If we actually see the rise of hostilities in space it would retard or perhaps even abruptly end all the exciting private space initiatives that are described in this book. One must always look to the future with optimism and recognize the sky is no longer the limit. Indeed the many riches of outer space now beckon.

Conclusions

The field of space research, applications and exploration is over sixty years old. There has been steady and consistent development in the areas of space technology, launch systems and practical applications of space systems over the past six decades. There have been many successful efforts to coordinate the uses of space during this time. There have been efforts to minimize interference, to make spacecraft networks safer, and to use spectrum efficiently, especially through the good offices of the International Telecommunication Union (ITU). There has also be useful coordination of efforts and improved forms of space governance achieved through the U. N. General Assembly, the U. N. Committee on the Peaceful Uses of Outer Space (COPUOS) and its Committees and Working Groups, plus the U. N. Office of Outer Space Affairs (OOSA). Specialized agencies of the United Nations that have contributed to this work have included not only the ITU but also the International Civil Aviation Organization (ICAO) (with regard to space traffic management) and the World Meteorological Organization (with regard to coordination of meteorological and remote-sensing/Earth observation satellites). Assistance has also been provided by U. N. Office of Disarmament Affairs (UNODA).

The original U. N. Outer Space Treaty and its four additional international agreements that were negotiated and agreed to in the 1960s and 1970s, set the basic ground rules for the peaceful uses of outer space. But since that time no major space agreements or treaties have been agreed to and ratified. This has led to a new era of agreement and cooperation that is sometimes referred to as 'soft law,' or the development of best practices, transparency and confidence-building measures (TCBM), and various efforts to create rules of the road in space or codes of conduct.

One of the true challenges that come from the Space 2.0 framework that is emerging is the lack of a legal framework for commercial space activities. Currently the activities of private enterprises in space keep expanding, and this commercial space activity is now on the order of three times the size of governmental and defense-related space activities. Despite this growth and despite the importance of commercial space activities all such effort remains explicitly under the licensing processes and responsibilities of nation states and in particular the registering launching state.

The space industry, as explained in the various chapters of this book, has been changed and, indeed, reinvented, especially in the last few years, but the legal framework and governance structure has not changed to reflect this new reality. The final step in this reinvention process seems clear. This changed state seems to cry out for some new governance systems and new approaches to regulating the future of space. On top of these new conditions and the expanding growth of commercial systems, there is the further complication that comes from the strategic and security requirements to limit the military uses of space systems that could also compromise the peaceful uses of outer space.

This issue also calls out for new ground rules and better means to restrict the possible deployment of space weapons in an iron-clad way so that no nation is tempted to violate the principles so well defined in the original Outer Space Treaty.

See Appendix C at the end of this book for a list of the U. N. Sustainable Development Goals and Space-Related Services.

References

1. 2017 State of the Satellite Industry Satellite Industry Association, June 2017. https://www.sia.org/annual-state-of-the-satellite-industry-reports/2017-sia-state-of-satellite-industry-report/
2. 2018 State of the Satellite Industry. Bryce Technology on behalf of the Satellite Industry Association, June 2018. https://brycetech.com/downloads/SIA_SSIR_2018.pdf
3. Arthur C. Clarke' three laws. The Arthur C. Clarke Foundation. http://www.clarkefoundation.org/about-sir-arthur/sir-arthurs-quotations/

Appendix A: Addendum to Chapter Three

Ten Possible Sources of Free Remote-Sensing Satellite Data

There are a growing number of commercial satellite systems that are structured to provide various types of Earth observation data for a wide variety of applications. There is also a wide range of governmentally sponsored sites around the world that are designed to help data analysts access information from remote-sensing satellites for a variety of purposes. Below are ten possible sites that might be used for free, after appropriate registration procedures. This list was initially prepared by Vinithra Rajendran. Further information can be found at http://geoawesomeness.com.

1. **GLOVIS**
   http://glovis.usgs.gov
   The USGS Global Visualization Viewer (GloVis) is one of the U. S. Geospatial Service's easier to use online search and order tools for selected satellite and aerial data. It is thus suited for initial users of Earth observation datasets.
   The products available for download include: Digital Orthophoto Quadrangle (DOQs), EO-1 ALI (Earth Observing-1 Advanced Land Imaging), EO-1 Hyperion (Earth Observing-1 Hyperion), Global Land Survey (GLS), Landsat 4-5 TM (L4-5 TM C1 Level-1), Landsat 7 ETM+ (L7 ETM+ C1 Level-1), Landsat 8 OLI/TIRS (L8 OLI/TIRS C1 Level-1), Sentinel-2, ISRO ResourceSAT 1 and 2 – AWIFS Sensor, ISRO ResourceSAT 1 and 2 – LISS-3 Sensor, GeoEye's OrbView-3 (OrbView-3). (Note: Registration is required)
2. **NASA Earth Observation (NEO)**
   NASA Earth observation allows access to more than 50 datasets on atmosphere, land, ocean, energy, environment, and other additional information. Some types of information may be available on a daily, weekly or monthly frequency. The datasets are available in the form of

J. N. Pelton, *Space 2.0*, Astronautical Engineering,
https://doi.org/10.1007/978-3-030-15281-9

JPEG, PNG, Google Earth and GeoTIFF formats. (Note: Registration is required.)

3. **USGS Earth Explorer**
http://arthexplorer.usgs.gov
In some ways the USGS Earth Explorer site is the most comprehensive website to provide free access to data from many different U. S. sources. Specially, there is not only a wide array of satellites, but also aerial images. It allows a wide range of search criteria. Further it allows the sequential arrangements of satellite imagery that is helpful in terms of downloading images and other useful data. USGS grants full access to NASA's Land Data Products and Services such as Hyperion, which represents useful hyperspectral data. It also provides access to radar data and MODIS & AVHRR land surface information. (Note: Registration is required.)
4. **ESA's Sentinel data**
http://cophub.copernicus.eu
The Copernicus Open Access Hub, which was previously known as Sentinels Scientific Data Hub, provides complete, free and open access to data from the Sentinel 1, 2 and 3 satellites as well as derived user products. These start from the In-Orbit Commissioning Review (IOCR). The data from the ESA Sentinel program is similar in a number of ways to that contained in the U. S. Geospatial Agency's Earth Explore site. (Note: Registration is required.)
5. **NASA Earth Data**
http://search.earthdata.nasa.gov
Earthdata Search, as of January 1, 2018, became the primary means for searching and discovering NASA Earth observing data. This new system seeks to support faster data searches and better search results for EOSDIS data users. Earthdata Search uses Client's natural language processing-enabled search tool to quickly narrow down to relevant collections. The Reverb data search and the discovery system are no longer operational.

   Earth Data includes data from a number of satellites such as NASA DC, GPS satellites, SMAP, JASON, METEOSAT, ALOS, TRMM, Aura, Aqua and others. (Note: Registration is required.)
6. **NOAA CLASS**
www.class.noaa.gov
NOAA (National Oceanic and Atmospheric Administration) CLASS website stands for the Comprehensive Large Array-data Stewardship System. It provides a useful online data library system of geographic data sets. The CLASS website draws from such sources as the U. S. Department of Defense (DOD) Polar-orbiting Operational Environmental Satellite (POES), Environmental Satellite (GOES), and other sources. (Note: Registration is required.)
7. **NOAA Digital Coast**
http://coast.noaa.gov/digitalcoast
If coastal data is your only requirement, then a portal that is focused on displaying coastal Earth Observation Imaging and operated by the U. S. Government's National Oceanic and Atmospheric Administration known as NOAA's Digital Coast can be a good source of data. In order to use this site one registers and then defines an area of interest and selects from the range

of free satellite imagery. There are a variety of datasets. These are organized by infrared, radar and true color composite to download. Apart from the coastal data you can also get imagery, land cover, elevation, socio-economic and benthic data. (Note: Registration is required.)

8. **IPPMUS Terra**
www.ipums.org/IPUMSTerra.shtml
Integrated Population and Environmental Data operates a website known as IPUMS Terra. This organization integrates population census data from around the world with global environmental data. This site allows users to obtain customized datasets that incorporate data from multiple sources in a single coherent structure. The country-specific data can be obtained from Terraclip featuring MODIS data. (Note: Registration is required.)
9. **LANCE**
http://earthdata.nasa.gov/lance or **http://lance.modaps.eosdis.nasa.gov**
The Land, Atmosphere Near real-time Capability for EOS (LANCE) is a NASA website and is a component of the NASA Earth Observing System Data and Information System (EOSDIS). This site is designed to support EOS application users interested in monitoring a wide variety of natural and man-made phenomena. This site, after registration, provides Near Real-Time (NRT) data and imagery from the AIRS, AMSR2, MISR, MLS, MODIS, OMI and VIIRS instruments. Data access in this makes information available much quicker than routine processing allows. Most data products are available within 3 hours from satellite observation. In addition NRT imagery is generally available 3-5 hours after observation. (Note: Registration is required.)
10. **VITO Vision**
https://vito.be/en
The Flemish Research Website VITO Vision (Vision on Technology for a Better World) provides information about various broad areas of vegetation from such sources as PROBA-V, SPOT-Vegetation and METOP. The low resolution data on this site provides broad sectors of vegetation patterns across Earth's surface. There is an easy-to-use interface and this low resolution satellite data is provided free. This type of data is a good for large-scale applications and environmental uses that do not need finer details. (Note: Registration is required.)

Please note that every portal makes registration mandatory to download data. These sites essentially provide access to U.S. Governmental and European Earth Observation web sites. There are other sites that are available, but the above are among the most accessible in terms of language and formatting.

Appendix B: Addendum to Chapter Five

Current, Past and Proposed Research Satellites for Climate-Change Purposes

Sat. Name	Status	Agency	Launch Date	Mission
ACRIMSAT	Failed December 2013	NASA	1999	Studied Sun's infrared to ultraviolet output.
AQUA	Active	NASA	2002	Carries six instruments to observe relation of Earth's systems: oceans, land, atmosphere and biosphere.
AQUARIUS	Active	NASA & Space Agency of Argentina	2010	Measures salt concentrations in ocean surface needed to understand heat transport and storage in the ocean.
AURA	Active	NASA	2004	Studies earth's ozone, air quality, and climate though observation of composition, chemistry, and dynamics of the atmosphere.
CALIPSO	Active	NASA	2006	Studies thickness of clouds and aerosols for understanding of how much air pollution is present and changes in compositions in the atmosphere.
Cloudsat	Active	NASA/ Canada	2006	Monitors the state of earth's atmosphere and weather through radar, which can be used to predict which clouds produce rain, observe snowfall, and monitor the moisture content of clouds.

(continued)

J. N. Pelton, *Space 2.0*, Astronautical Engineering,
https://doi.org/10.1007/978-3-030-15281-9

(continued)

Sat. Name	Status	Agency	Launch Date	Mission
Deepspace Climate Observatory	Active	NASA	2015	To study the Sun-lit side of Earth from the L1 Lagrange point
EarthCARE	Active	ESA/JAXA	2013	EarthCARE – Study of clouds and aerosols.
Earth Observing-1 (NMP)	Active	NASA	2001	Carryiesland-imaging technology, used to demonstrate new instruments and spacecraft systems for future missions.
Global Precipitation Measurement	Active	NASA	2014	Studies global precipitation.
GLORY	Launch failure	NASA	2011	Studies aerosols, including black carbon, in addition to solar irradiance for the long-term effects.
GOES-1 M	Active	NASA	2001	Monitors and forecasts weather for NOAA
GRACE	Active	NASA and German space agency DLR	2002	Observes and measures earth's gravitational field, which may help determining the shape and composition of the planet's distribution of water and ice.
ICESat	Active	NASA	2003	Keeps track of size and thickness of earth's ice sheets.
Jason-1, 2, &3	Active	NASA	2001, 2008, 2016	They use a radar altimeter to monitor ocean surface height.
LAGEOS 1&2	Active	NASA	1976	LAGEOS 1 launched in 1976, LAGEOS 2, launched in 1992 used for orbiting benchmark for geodynamical studies.
Landsat 7	Active	NASA	1999	Takes digital images of earth's coastal areas with global coverage on a seasonal basis.
Landsat 8	Active	NASA	2013	Takes digital images of earth's coastal areas with global coverage on a seasonal basis.
QuikSCAT	Active	NASA	1997	Monitors weather using bursts of microwaves which measure wind speeds.
SEASTAR (SEAWIFS)	Active	NASA	1997	Designed to monitor the color of earth's oceans.
SMAP	Active, but with partial failure	NASA	2015	Measures soil moisture and its freeze/thaw state, which helps understanding of processes that link water, energy, and carbon cycles. Radar failed, but radiometer is working.

(continued)

(continued)

Sat. Name	Status	Agency	Launch Date	Mission
SORCE	Active	NASA	2003	Monitors total output from the sun for understanding of earth's absorption of radiation energy.
TERRA	Active	NASA/ Canada/ Japan	1999	Carries five instruments to observe the state of the atmosphere, land, and oceans, as well as their interactions with solar radiation
TRMM	Active	NASA	1997	Carries give instruments which uses radar and sensors of visible infrared light to closely monitor precipitation.
CLARREO	Proposed	NASA		Measures spectrally resolved Earth's reflectance and emitted radiation, and radio occultation derived refractivity;
ICESat-II	Active	NASA	2015	Measure ice sheet height changes for climate change diagnoses.
DESDynI	Proposed	NASA		Measures surface and ice sheet deformation to determine natural hazards of climate.
HyspIRI	Proposed	NASA		Monitors land surface composition for agriculture and mineral characterization for ecosystem health.
ASCENDS	Proposed	NASA		Measures the number density of CO_2 in a column of beneath the craft in addition to ambient temperature and pressure.
SWOT	Proposed	NASA		Tracks ocean, lake, river levls.
GEO-CAPE	Proposed	NASA		Monitors atmospheric gas columns for air-quality forecasts.
ACE	Proposed	NASA		Using lidar, creates aerosol and cloud profiles.
LIST	Proposed	NASA		Measure surface topography to look for landslide hazard and water runoffs.

Appendix C: Addendum to Chapter Fourteen

Following are the U. N. Sustainable Development Goals and Space-Related Services and how they are supported by the various space industries.

The United Nations 17 Sustainable Development Goals and Space-Related Activities		
Goal	**Description**	**Illustrative Space-Based Activity**
1	**End poverty in all its forms everywhere**	• Telecommunication support to banks and financial institutions in more outreach programs and allow better ease-of-access to financial institutions. • Monitoring, evaluating, and making satellite data available to international development banks, which invest in places with least development. • EO for mapping agriculture and livestock, enabling the sustainable utilization of land for crops, grazing, and water.
2	**End hunger, achieve food security and improved nutrition, and promote sustainable agriculture**	• Remote sensing data strengthens the global monitoring of agriculture and health of livestock. • Weather forecasting helps with reliable, accurate, timely, and sustained crop monitoring information and yield forecasts that significantly contribute to global food security.
3	**Ensure healthy lives and promote well-being for all at all ages**	• Tele-medicine helps in remote diagnosis utilizing satellite communications in remote areas, mapping and assessing areas with epidemics. Orbital health research contributes to the development of innovative solutions for promoting healthy lives and well-being. • PNT assists in tracking and monitoring patients. For example, elderly people can bear devices that detect a fall or collapse and can immediately notify

(continued)

J. N. Pelton, *Space 2.0*, Astronautical Engineering,
https://doi.org/10.1007/978-3-030-15281-9

(continued)

The United Nations 17 Sustainable Development Goals and Space-Related Activities		
Goal	**Description**	**Illustrative Space-Based Activity**
4	**Ensure inclusive and equitable quality education and promote lifelong learning opportunities for all.**	• Tele-education helps reach rural areas and brings access to education in more economical ways compared to setting up full-scale infrastructure and bringing expertise to such areas. • Access to flying experiments in orbit bring special environment conditions (zero or micro gravity) that are hard to create on Earth and help study their effects.
5	**Achieve gender equality and empower all women and girls**	• Space industry, with its increasing workforce, provides employment opportunity to women. Furthermore, greater access to information on global level increases awareness of gender equality and rights. Again, information of and increased access to women hygiene, health, and maternity help reduce death rates in women.
6	**Ensure availability and sustainable management of water and sanitation for all**	• EO methods, like the digital elevation modelling (DEM), are the representation of a terrain (like Earth) showing elevation values of a topographic surface. This helps in modeling water flow for hydrology – a study focused on the movement, distribution, and quality of water on Earth.
7	**Ensure access to affordable, reliable, sustainable, and modern energy for all**	• Spin-offs of energy research projects in space industry. • Space-based solar power aimed at collecting solar power in space and transmitting it back to Earth.
8	**Promote sustained, inclusive, & sustainable economic growth, full & productive employment and decent work for all**	• Space technology significantly contributes to revenue generation and sustainable economy. Programs like the ESA's Copernicus are estimated to return EUR 10 for every EUR 1 invested. Similarly accepted estimate for NASA programs is US$ 7-10 for every US$ 1 spent. • With growing space utilization and investments, the workforce is likely to increase manifold.
9	**Build resilient infrastructure, promote inclusive and sustainable industrialization, and foster innovation**	• With the EO system providing data on environmental impacts, climate change, and weather patterns, it is easy to plan and manage industrial activities in a sustainable manner. Research and development in the space industry help promoting innovation that benefits space exploration and the spin-offs bring great benefits to the global community.
10	**Reduce inequality within and among countries**	• An increasing number of States assisting in launch activity, commercially and cooperatively, reduces inequality and promotes accessibility to space. Particularly, the trending small satellite use and piggybacking of payloads can serve to be great means toward the democratization of space.

(continued)

(continued)

The United Nations 17 Sustainable Development Goals and Space-Related Activities		
Goal	**Description**	**Illustrative Space-Based Activity**
11	**Make cities and human settlements inclusive, safe, resilient, and sustainable**	• Urban mapping with the help of PNT and EO data can help plan and manage efficient sustainable urban development. With hand-held PNT devices and surveillance through satellites, great extent of safety can be achieved.
12	**Ensure sustainable consumption and production patterns**	• With the help of satellites, more efficiency can be achieved in predicting production patterns and, with the use of big data, consumption can be monitored with high accuracy. Particularly, in the agricultural industry, farmers can be notified of demand, supply, and rates along with extremely important land-use data for crop management.
13	**Take urgent action to combat climate change and its impacts**	• The EO systems help in monitoring land use data through which deforestation, desertification, and watershed management is possible with high accuracy and efficiency. Monitoring the weather patterns, climate change, and carbon emissions over a time in any region on the Earth helps in combating climate change.
14	**Conserve and sustainably use the oceans, seas, and marine resources for sustainable development**	• The EO and PNT systems help in monitoring oceans and marine resources, and also make marine navigation more accurate. Marine biodiversity can also be tracked and monitored through satellite data.
15	**Protect, restore, and promote sustainable use of terrestrial ecosystems, sustainably manage forests, combat desertification, and halt and reverse land degradation and halt biodiversity loss**	• The EO systems help in monitoring land-use data through which biodiversity management, combating deforestation, combating desertification, and watershed management are possible with high accuracy and efficiency.
16	**Promote peaceful and inclusive societies for sustainable development, provide access to justice for all, and build effective, accountable and, inclusive institutions at all levels**	• With the EO and PNT systems, great levels of transparency can be brought to any political, economic or societal system. More transparency helps build just, equitable, and more inclusive institutions.
17	**Strengthen the means of implementation and revitalize the global partnership for sustainable development**	• Space activities, since their inception, have seen some of the largest global scientific and economic cooperation. This will continue to grow, and it is guaranteed that global partnerships for sustainable developments will strengthen space exploration and use.

Appendix D: Space Policy Directives

Space Directive 1: Presidential Memorandum Issued on December 11, 2017

MEMORANDUM FOR THE VICE PRESIDENT
THE SECRETARY OF STATE
THE SECRETARY OF DEFENSE
THE SECRETARY OF COMMERCE
THE SECRETARY OF TRANSPORTATION
THE SECRETARY OF HOMELAND SECURITY
THE DIRECTOR OF NATIONAL INTELLIGENCE
THE DIRECTOR OF THE OFFICE OF MANAGEMENT AND BUDGET
THE ASSISTANT TO THE PRESIDENT FOR NATIONAL SECURITY AFFAIRS
THE ADMINISTRATOR OF THE NATIONAL AERONAUTICS AND SPACE ADMINISTRATION
THE DIRECTOR OF THE OFFICE OF SCIENCE AND TECHNOLOGY POLICY
THE ASSISTANT TO THE PRESIDENT FOR HOMELAND SECURITY AND COUNTERTERRORISM
THE CHAIRMAN OF THE JOINT CHIEFS OF STAFF

SUBJECT: Reinvigorating America's Human Space Exploration Program

Sec. 1. Amendment to Presidential Policy Directive-4

Presidential Policy Directive-4 of June 28, 2010 (National Space Policy) is amended as follows:

The paragraph beginning "Set far-reaching exploration milestones" is deleted and replaced with the following:

"Lead an innovative and sustainable program of exploration with commercial and international partners to enable human expansion across the solar system and to bring back to Earth new knowledge and opportunities. Beginning with missions beyond low-Earth orbit, the United States will lead the return of humans to the Moon for long-term

J. N. Pelton, *Space 2.0*, Astronautical Engineering,
https://doi.org/10.1007/978-3-030-15281-9

exploration and utilization, followed by human missions to Mars and other destinations;".

Sec. 2. General Provisions

(a) Nothing in this memorandum shall be construed to impair or otherwise affect:
  (i) the authority granted by law to an executive department or agency, or the head thereof; or
  (ii) the functions of the Director of the Office of Management and Budget relating to budgetary, administrative, or legislative proposals.
(b) This memorandum shall be implemented consistent with applicable law and subject to the availability of appropriations.
(c) This memorandum is not intended to, and does not, create any right or benefit, substantive or procedural, enforceable at law or in equity by any party against the United States, its departments, agencies, or entities, its officers, employees, or agents, or any other person.
(d) This memorandum shall be published in the Federal Register.

DONALD J. TRUMP
White House Logo
The White House

Space Policy Directive-2: Presidential Memorandum Issued on May 24, 2018

MEMORANDUM FOR THE VICE PRESIDENT
THE SECRETARY OF STATE
THE SECRETARY OF DEFENSE
THE SECRETARY OF COMMERCE
THE SECRETARY OF TRANSPORTATION
THE SECRETARY OF HOMELAND SECURITY
THE SECRETARY OF LABOR
THE DIRECTOR OF NATIONAL INTELLIGENCE
THE DIRECTOR OF THE OFFICE OF MANAGEMENT AND BUDGET
THE ASSISTANT TO THE PRESIDENT FOR NATIONAL SECURITY AFFAIRS
THE ADMINISTRATOR OF THE NATIONAL AERONAUTICS AND SPACE ADMINISTRATION
THE DIRECTOR OF THE OFFICE OF SCIENCE AND TECHNOLOGY POLICY
THE ASSISTANT TO THE PRESIDENT FOR HOMELAND SECURITY AND COUNTERTERRORISM
THE CHAIRMAN OF THE JOINT CHIEFS OF STAFF

SUBJECT: Streamlining Regulations on Space Infrastructure and Technology

Sec. 1. Policy

It is the policy of the executive branch to be prudent and responsible when spending taxpayer funds, and to recognize how government actions, including Federal regulations, affect private resources. It is therefore important that regulations adopted and enforced by the executive branch promote economic growth; minimize uncertainty for taxpayers, investors, and private industry; protect national security, public-safety, and foreign policy interests; and encourage American leadership in space commerce.

Sec. 2. Launch and Re-entry Licensing

(a) No later than February 1, 2019, the Secretary of Transportation shall review regulations adopted by the Department of Transportation that provide for and govern licensing of commercial space flight launch and re-entry for consistency with the policy set forth in section 1 of this memorandum and shall rescind or revise those regulations, or publish for notice and comment proposed rules rescinding or revising those regulations, as appropriate and consistent with applicable law.

(b) Consistent with the policy set forth in section 1 of this memorandum, the Secretary of Transportation shall consider the following:
   (i) requiring a single license for all types of commercial space flight launch and re-entry operations; and
   (ii) replacing prescriptive requirements in the commercial space flight launch and re-entry licensing process with performance-based criteria.

(c) In carrying out the review required by subsection (a) of this section, the Secretary of Transportation shall coordinate with the members of the National Space Council.

(d) The Secretary of Defense, the Secretary of Transportation, and the Administrator of the National Aeronautics and Space Administration shall coordinate to examine all existing U.S. Government requirements, standards, and policies associated with commercial space flight launch and re entry operations from Federal launch ranges and, as appropriate and consistent with applicable law, to minimize those requirements, except those necessary to protect public safety and national security, that would conflict with the efforts of the Secretary of Transportation in implementing the Secretary's responsibilities under this section.

Sec. 3. Commercial Remote Sensing

(a) Within 90 days of the date of this memorandum, the Secretary of Commerce shall review the regulations adopted by the Department of Commerce under Title II of the Land Remote Sensing Policy Act of 1992 (51 U.S.C. 60101 et seq.) for consistency with the policy set forth in section 1 of this memorandum and shall rescind or revise those regulations, or publish for notice and comment proposed rules rescinding or revising those regulations, as appropriate and consistent with applicable law.

(b) In carrying out the review required by subsection (a) of this section, the Secretary of Commerce shall coordinate with the Secretary of State, the Secretary of Defense, the Administrator of the National Aeronautics and Space Administration, and, as appropriate, the Chairman of the Federal Communications Commission.

(c) Within 120 days of the date of the completion of the review required by subsection (a) of this section, the Secretary of Commerce, in coordination with the Secretary of State and the Secretary of Defense, shall transmit to the Director of the Office of Management and Budget a legislative proposal to encourage expansion of the licensing of commercial remote sensing activities. That proposal shall be consistent with the policy set forth in section 1 of this memorandum.

Sec. 4. Reorganization of the Department of Commerce

(a) To the extent permitted by law, the Secretary of Commerce shall consolidate in the Office of the Secretary of Commerce the responsibilities of the Department of Commerce with respect to the Department's regulation of commercial space flight activities.
(b) Within 30 days of the date of this memorandum, the Secretary of Commerce shall transmit to the Director of the Office of Management and Budget a legislative proposal to create within the Department of Commerce an entity with primary responsibility for administering the Department's regulation of commercial space flight activities.

Sec. 5. Radio Frequency Spectrum

(a) The Secretary of Commerce, in coordination with the Director of the Office of Science and Technology Policy, shall work with the Federal Communications Commission to ensure that Federal Government activities related to radio frequency spectrum are, to the extent permitted by law, consistent with the policy set forth in section 1 of this memorandum.
(b) Within 120 days of the date of this memorandum, the Secretary of Commerce and the Director of the Office of Science and Technology Policy, in consultation with the Chairman of the Federal Communications Commission, and in coordination with the members of the National Space Council, shall provide to the President, through the Executive Secretary of the National Space Council, a report on improving the global competitiveness of the United States space sector through radio frequency spectrum policies, regulation, and United States activities at the International Telecommunication Union and other multilateral forums.

Sec. 6. Review of Export Licensing Regulations

The Executive Secretary of the National Space Council, in coordination with the members of the National Space Council, shall:

(a) initiate a review of export licensing regulations affecting commercial space flight activity;
(b) develop recommendations to revise such regulations consistent with the policy set forth in section 1 of this memorandum and with applicable law; and
(c) submit such recommendations to the President, through the Vice President, no later than 180 days from the date of this memorandum.

Sec. 7. General Provisions

(a) Nothing in this memorandum shall be construed to impair or otherwise affect:
  (i) the authority granted by law to an executive department or agency, or the head thereof; or
  (ii) the functions of the Director of the Office of Management and Budget relating to budgetary, administrative, or legislative proposals.
(b) This memorandum shall be implemented consistent with applicable law and subject to the availability of appropriations.
(c) This memorandum is not intended to, and does not, create any right or benefit, substantive or procedural, enforceable at law or in equity by any party against the United States, its departments, agencies, or entities, its officers, employees, or agents, or any other person.
(d) The Secretary of Transportation is authorized and directed to publish this memorandum in the Federal Register.

https://www.whitehouse.gov/presidential-actions/space-policy-directive-2-streamlining-regulations-commercial-use-space/

Space Policy Directive-3: Presidential Memorandum Issued on June 18, 2018

MEMORANDUM FOR THE VICE PRESIDENT
THE SECRETARY OF STATE
THE SECRETARY OF DEFENSE
THE SECRETARY OF COMMERCE
THE SECRETARY OF TRANSPORTATION
THE SECRETARY OF HOMELAND SECURITY
THE DIRECTOR OF NATIONAL INTELLIGENCE
THE DIRECTOR OF THE OFFICE OF MANAGEMENT AND BUDGET
THE ASSISTANT TO THE PRESIDENT FOR NATIONAL SECURITY AFFAIRS
THE ADMINISTRATOR OF THE NATIONAL AERONAUTICS AND SPACE ADMINISTRATION
THE DIRECTOR OF THE OFFICE OF SCIENCE AND TECHNOLOGY POLICY
THE DEPUTY ASSISTANT TO THE PRESIDENT FOR HOMELAND SECURITY AND COUNTERTERRORISM
THE CHAIRMAN OF THE JOINT CHIEFS OF STAFF

SUBJECT: National Space Traffic Management Policy

Sec. 1. Policy

For decades, the United States has effectively reaped the benefits of operating in space to enhance our national security, civil, and commercial sectors. Our society now depends on space technologies and space-based capabilities for communications, navigation, weather forecasting, and much more. Given the significance of space activities, the United States considers the continued unfettered access to and freedom to operate in space of vital interest to advance the security, economic prosperity, and scientific knowledge of the Nation.

Today, space is becoming increasingly congested and contested, and that trend presents challenges for the safety, stability, and sustainability of U. S. space operations. Already, the Department of Defense (DOD) tracks over 20,000 objects in space, and that number will increase dramatically as new, more capable sensors come online and are able to detect smaller objects. DOD publishes a catalog of space objects and makes notifications of potential conjunctions (that is, two or more objects coming together at the same or nearly the same point in time and space). As the number of space objects increases, however, this limited traffic management activity and architecture will become inadequate. At the same time, the contested nature of space is increasing the demand for DOD focus on protecting and defending U.S. space assets and interests.

The future space operating environment will also be shaped by a significant increase in the volume and diversity of commercial activity in space. Emerging commercial ventures such as satellite servicing, debris removal, in-space manufacturing, and tourism, as well as new technologies enabling small satellites and very large constellations of satellites, are increasingly outpacing efforts to develop and implement government policies and processes to address these new activities.

To maintain U.S. leadership in space, we must develop a new approach to space traffic management (STM) that addresses current and future operational risks. This new approach must set priorities for space situational awareness (SSA) and STM innovation in science and technology (S&T), incorporate national security considerations, encourage growth of the U.S. commercial space sector, establish an updated STM architecture, and promote space safety standards and best practices across the international community.

The United States recognizes that spaceflight safety is a global challenge and will continue to encourage safe and responsible behavior in space while emphasizing the need for international transparency and STM data sharing. Through this national policy for STM and other national space strategies and policies, the United States will enhance safety and ensure continued leadership, preeminence, and freedom of action in space.

Sec. 2. Definitions

For the purposes of this memorandum, the following definitions shall apply:

(a) Space Situational Awareness shall mean the knowledge and characterization of space objects and their operational environment to support safe, stable, and sustainable space activities.
(b) Space Traffic Management shall mean the planning, coordination, and on-orbit synchronization of activities to enhance the safety, stability, and sustainability of operations in the space environment.
(c) Orbital debris, or space debris, shall mean any human-made space object orbiting Earth that no longer serves any useful purpose.

Sec. 3. Principles

The United States recognizes, and encourages other nations to recognize, the following principles:

(a) Safety, stability, and operational sustainability are foundational to space activities, including commercial, civil, and national security activities. It is a shared interest and responsibility of all spacefaring nations to create the conditions for a safe, stable, and operationally sustainable space environment.

(b) Timely and actionable SSA data and STM services are essential to space activities. Consistent with national security constraints, basic U.S. Government-derived SSA data and basic STM services should be available free of direct user fees.

(c) Orbital debris presents a growing threat to space operations. Debris mitigation guidelines, standards, and policies should be revised periodically, enforced domestically, and adopted internationally to mitigate the operational effects of orbital debris.

(d) A STM framework consisting of best practices, technical guidelines, safety standards, behavioral norms, pre-launch risk assessments, and on-orbit collision avoidance services is essential to preserve the space operational environment.

Sec. 4. Goals

Consistent with the principles listed in section 3 of this memorandum, the United States should continue to lead the world in creating the conditions for a safe, stable, and operationally sustainable space environment. Toward this end, executive departments and agencies (agencies) shall pursue the following goals as required in section 6 of this memorandum:

(a) Advance SSA and STM Science and Technology. The United States should continue to engage in and enable S&T research and development to support the practical applications of SSA and STM. These activities include improving fundamental knowledge of the space environment, such as the characterization of small debris, advancing the S&T of critical SSA inputs such as observational data, algorithms, and models necessary to improve SSA capabilities, and developing new hardware and software to support data processing and observations.

(b) Mitigate the effect of orbital debris on space activities. The volume and location of orbital debris are growing threats to space activities. It is in the interest of all to minimize new debris and mitigate effects of existing debris. This fact, along with increasing numbers of active satellites, highlights the need to update existing orbital debris mitigation guidelines and practices to enable more efficient and effective compliance, and establish standards that can be adopted internationally. These trends also highlight the need to establish satellite safety design guidelines and best practices.

(c) Encourage and facilitate U.S. commercial leadership in S&T, SSA, and STM. Fostering continued growth and innovation in the U.S. commercial space sector, which includes S&T, SSA, and STM activities, is in the national interest of the United States. To achieve this goal, the U.S. Government should streamline processes and reduce regulatory burdens that could inhibit commercial sector growth and innovation, enabling the U.S. commercial sector to continue to lead the world in

STM-related technologies, goods, data, and services on the international market.

(d) Provide U.S. Government-supported basic SSA data and basic STM services to the public. The United States should continue to make available basic SSA data and basic STM services (including conjunction and reentry notifications) free of direct user fees while supporting new opportunities for U.S. commercial and non-profit SSA data and STM services.

(e) Improve SSA data interoperability and enable greater SSA data sharing. SSA data must be timely and accurate. It is in the national interest of the United States to improve SSA data interoperability and enable greater SSA data sharing among all space operators, consistent with national security constraints. The United States should seek to lead the world in the development of improved SSA data standards and information sharing.

(f) Develop STM standards and best practices. As the leader in space, the United States supports the development of operational standards and best practices to promote safe and responsible behavior in space. A critical first step in carrying out that goal is to develop U.S.-led minimum safety standards and best practices to coordinate space traffic. U.S. regulatory agencies should, as appropriate, adopt these standards and best practices in domestic regulatory frameworks and use them to inform and help shape international consensus practices and standards.

(g) Prevent unintentional radio frequency (RF) interference. Growing orbital congestion is increasing the risk to U. S. space assets from unintentional RF interference. The United States should continue to improve policies, processes, and technologies for spectrum use (including allocations and licensing) to address these challenges and ensure appropriate spectrum use for current and future operations.

(h) Improve the U. S. domestic space object registry. Transparency and data sharing are essential to safe, stable, and sustainable space operations. Consistent with national security constraints, the United States should streamline the interagency process to ensure accurate and timely registration submissions to the United Nations (UN), in accordance with our international obligations under the Convention on Registration of Objects Launched into Outer Space.

(i) Develop policies and regulations for future U.S. orbital operations. Increasing congestion in key orbits and maneuver-based missions such as servicing, survey, and assembly will drive the need for policy development for national security, civil, and commercial sector space activities. Consistent with U. S. law and international obligations, the United States should regularly assess existing guidelines for non-government orbital activities, and maintain a timely and responsive regulatory environment for licensing these activities.

Sec. 5. Guidelines

In pursuit of the principles and goals of this policy, agencies should observe the following guidelines:

(a) Managing the Integrity of the Space Operating Environment.

(i) Improving SSA coverage and accuracy. Timely, accurate, and actionable data are essential for effective SSA and STM. The United States should seek to minimize deficiencies in SSA capability, particularly coverage in regions with limited sensor availability and sensitivity in detection of small debris, through SSA data sharing, the purchase of SSA data, or the provision of new sensors.

(ii) New U. S. sensors are expected to reveal a substantially greater volume of debris and improve our understanding of space object size distributions in various regions of space. However, very small debris may not be sufficiently tracked to enable or justify actionable collision avoidance decisions. As a result, close conjunctions and even collisions with unknown objects are possible, and satellite operators often lack sufficient insight to assess their level of risk when making maneuvering decisions. The United States should develop better tracking capabilities, and new means to catalog such debris, and establish a quality threshold for actionable collision avoidance warning to minimize false alarms.

(iii) Through both Government and commercial sector S&T investment, the United States should advance concepts and capabilities to improve SSA in support of debris mitigation and collision avoidance decisions.

(iv) Establishing an Open Architecture SSA Data Repository. Accurate and timely tracking of objects orbiting Earth is essential to preserving the safety of space activities for all. Consistent with section 2274 of title 10, United States Code, a basic level of SSA data in the form of the publicly releasable portion of the DoD catalog is and should continue to be provided free of direct user fees. As additional sources of space tracking data become available, the United States has the opportunity to incorporate civil, commercial, international, and other available data to allow users to enhance and refine this service. To facilitate greater data sharing with satellite operators and enable the commercial development of enhanced space safety services, the United States must develop the standards and protocols for creation of an open architecture data repository. The essential features of this repository would include:

- Data integrity measures to ensure data accuracy and availability;
- Data standards to ensure sufficient quality from diverse sources;

- Measures to safeguard proprietary or sensitive data, including national security information;
- The inclusion of satellite owner-operator ephemerides to inform orbital location and planned maneuvers; and
- Standardized formats to enable development of applications to leverage the data.

(v) To facilitate this enhanced data sharing, and in recognition of the need for DoD to focus on maintaining access to and freedom of action in space, a civil agency should, consistent with applicable law, be responsible for the publicly releasable portion of the DoD catalog and for administering an open architecture data repository. The Department of Commerce should be that civil agency.

(vi) Mitigating Orbital Debris. It is in the interest of all space operators to minimize the creation of new orbital debris. Rapid international expansion of space operations and greater diversity of missions have rendered the current U.S. Government Orbital Debris Mitigation Standard Practices (ODMSP) inadequate to control the growth of orbital debris. These standard practices should be updated to address current and future space operating environments.

The United States should develop a new protocol of standard practices to set broader expectations of safe space operations in the 21st century. This protocol should begin with updated ODMSP, but also incorporate sections to address operating practices for large constellations, rendezvous and proximity operations, small satellites, and other classes of space operations. These overarching practices will provide an avenue to promote efficient and effective space safety practices with U.S. industry and internationally.

The United States should pursue active debris removal as a necessary long-term approach to ensure the safety of flight operations in key orbital regimes. This effort should not detract from continuing to advance international protocols for debris mitigation associated with current programs.

(b) Operating in a Congested Space Environment

(i) Minimum Safety Standards and Best Practices. The creation of minimum standards for safe operation and debris mitigation derived in part from the U. S. Government ODMSP, but incorporating other standards and best practices, will best ensure the safe operation of U. S. space activities. These safety guidelines should consider maneuverability, tracking, reliability, and disposal.

The United States should eventually incorporate appropriate standards and best practices into Federal law and regulation through appropriate rulemaking or licensing actions. These guidelines should encompass protocols for all stages of satellite operation from design through end-of-life.

Satellite and constellation owners should participate in a pre-launch certification process that should, at a minimum, consider the following factors:

- Coordination of orbit utilization to prevent conjunctions;
- Constellation owner-operators' management of self-conjunctions;
- Owner-operator notification of planned maneuvers and sharing of satellite orbital location data;
- On-orbit tracking aids, including beacons or sensing enhancements, if such systems are needed;
- Encryption of satellite command and control links and data protection measures for ground site operations;
- Appropriate minimum reliability based on type of mission and phase of operations;
- Effect on the national security or foreign policy interests of the United States, or international obligations; and
- Self-disposal upon the conclusion of operational lifetime, or owner-operator provision for disposal using active debris removal methods.

(ii) On-Orbit Collision Avoidance Support Service. Timely warning of potential collisions is essential to preserving the safety of space activities for all. Basic collision avoidance information services are and should continue to be provided free of direct user fees. The imminent activation of more sensitive tracking sensors is expected to reveal a significantly greater population of the existing orbital debris background as well as provide an improved ability to track currently catalogued objects. Current and future satellites, including large constellations of satellites, will operate in a debris environment much denser than presently tracked. Preventing on-orbit collisions in this environment requires an information service that shares catalog data, predicts close approaches, and provides actionable warnings to satellite operators. The service should provide data to allow operators to assess proposed maneuvers to reduce risk. To provide on-orbit collision avoidance, the United States should:

- Provide services based on a continuously updated catalog of satellite tracking data;
- Utilize automated processes for collision avoidance;
- Provide actionable and timely conjunction assessments; and
- Provide data to operators to enable assessment of maneuver plans.

To ensure safe coordination of space traffic in this future operating environment, and in recognition of the need for DoD to focus on maintaining access to and freedom of action in space, a civil agency should be the focal point for this collision avoidance support service. The Department of Commerce should be that civil agency

(c) Strategies for Space Traffic Management in a Global Context.

(i) Protocols to Prevent Orbital Conjunctions. As increased satellite operations make lower Earth orbits more congested, the United States should develop a set of standard techniques for mitigating the collision risk of increasingly congested orbits, particularly for large constellations. Appropriate methods, which may include licensing assigned volumes for constellation operation and establishing processes for satellites passing through the volumes, are needed.

The United States should explore strategies that will lead to the establishment of common global best practices, including:

- A common process addressing the volume of space used by a large constellation, particularly in close proximity to an existing constellation;
- A common process by which individual spacecraft may transit volumes used by existing satellites or constellations; and
- A set of best practices for the owner-operators of utilized volumes to minimize the long-term effects of constellation operations on the space environment (including the proper disposal of satellites, reliability standards, and effective collision avoidance).

(ii) Radio Frequency Spectrum and Interference Protection. Space traffic and RF spectrum use have traditionally been independently managed processes. Increased congestion in key orbital regimes creates a need for improved and increasingly dynamic methods to coordinate activities in both the physical and spectral domains, and may introduce new interdependencies. U.S. Government efforts in STM should address the following spectrum management considerations:

- Where appropriate, verify consistency between policy and existing national and international regulations and goals regarding global access to, and operation in, the RF spectrum for space services;
- Investigate the advantages of addressing spectrum in conjunction with the development of STM systems, standards, and best practices;

- Promote flexible spectrum use and investigate emerging technologies for potential use by space systems; and
- Ensure spectrum-dependent STM components, such as inter-satellite safety communications and active debris removal systems, can successfully access the required spectrum necessary to their missions.

(iii) Global Engagement. In its role as a major spacefaring nation, the United States should continue to develop and promote a range of norms of behavior, best practices, and standards for safe operations in space to minimize the space debris environment and promote data sharing and coordination of space activities. It is essential that other spacefaring nations also adopt best practices for the common good of all spacefaring states. The United States should encourage the adoption of new norms of behavior and best practices for space operations by the international community through bilateral and multilateral discussions with other spacefaring nations, and through U.S. participation in various organizations such as the Inter-Agency Space Debris Coordination Committee, International Standards Organization, Consultative Committee for Space Data Systems, and UN Committee on the Peaceful Uses of Outer Space.

Sec. 6. Roles and Responsibilities

In furtherance of the goals described in section 4 and the guidelines described in section 5 of this memorandum, agencies shall carry out the following roles and responsibilities:

(a) Advance SSA and STM S&T. Members of the National Space Council, or their delegees, shall coordinate, prioritize, and advocate for S&T, SSA, and STM, as appropriate, as it relates to their respective missions. They should seek opportunities to engage with the commercial sector and academia in pursuit of this goal.

(b) Mitigate the Effect of Orbital Debris on Space Activities.

(i) The Administrator of the National Aeronautics and Space Administration (NASA Administrator), in coordination with the Secretaries of State, Defense, Commerce, and Transportation, and the Director of National Intelligence, and in consultation with the Chairman of the Federal Communications Commission (FCC), shall lead efforts to update the U. S. Orbital Debris Mitigation Standard Practices and establish new guidelines for satellite design and operation, as appropriate and consistent with applicable law.

(ii) The Secretaries of Commerce and Transportation, in consultation with the Chairman of the FCC, will assess the suitability of incorporating these updated standards and best practices into their respective licensing processes, as appropriate and consistent with applicable law.

(c) Encourage and Facilitate U.S. Commercial Leadership in S&T, SSA, and STM. The Secretary of Commerce, in coordination with the Secretaries of Defense and Transportation, and the NASA Administrator, shall lead efforts to encourage and facilitate continued U.S. commercial leadership in SSA, STM, and related S&T.

(d) Provide U.S. Government-Derived Basic SSA Data and Basic STM Services to the Public.

(i) The Secretaries of Defense and Commerce, in coordination with the Secretaries of State and Transportation, the NASA Administrator, and the Director of National Intelligence, should cooperatively develop a plan for providing basic SSA data and basic STM services either directly or through a partnership with industry or academia, consistent with the guidelines of sections 5(a)(ii) and 5(b)(ii) of this memorandum.

(ii) The Secretary of Defense shall maintain the authoritative catalog of space objects.

(iii) The Secretaries of Defense and Commerce shall assess whether statutory and regulatory changes are necessary to effect the plan developed under subsection (d)(i) of this section, and shall pursue such changes, along with any other needed changes, as appropriate.

(e) Improve SSA Data Interoperability and Enable Greater SSA Data Sharing.

(i) The Secretary of Commerce, in coordination with the Secretaries of State, Defense, and Transportation, the NASA Administrator, and the Director of National Intelligence, shall develop standards and protocols for creation of an open architecture data repository to improve SSA data interoperability and enable greater SSA data sharing.

(ii) The Secretary of Commerce shall develop options, either in-house or through partnerships with industry or academia, assessing both the technical and economic feasibility of establishing such a repository.

(iii) The Secretary of Defense shall ensure that release of data regarding national security activities to any person or entity with access to the repository is consistent with national security interests.

(f) Develop Space Traffic Standards and Best Practices. The Secretaries of Defense, Commerce, and Transportation, in coordination with the Secretary of State, the NASA Administrator, and the Director of National Intelligence, and in consultation with the Chairman of the FCC, shall develop space traffic standards and best practices, including technical guidelines, minimum safety standards, behavioral norms, and orbital conjunction prevention protocols related to pre-launch risk assessment and on-orbit collision avoidance support services.

(g) Prevent Unintentional Radio Frequency Interference. The Secretaries of Commerce and Transportation, in coordination with the Secretaries of State and Defense, the NASA Administrator, and the Director of National Intelligence, and in consultation with the Chairman of the FCC, shall coordinate to mitigate the risk of harmful interference and promptly address any harmful interference that may occur.

(h) Improve the U.S. Domestic Space Object Registry. The Secretary of State, in coordination with the Secretaries of Defense, Commerce, and Transportation, the NASA Administrator, and the Director of National Intelligence, and in consultation with the Chairman of the FCC, shall lead U.S. Government efforts on international engagement related to international transparency and space object registry on SSA and STM issues.

(i) Develop Policies and Regulations for Future U.S. Orbital Operations. The Secretaries of Defense, Commerce, and Transportation, in coordination with the Secretary of State, the NASA Administrator, and the Director of National Intelligence, shall regularly evaluate emerging trends in space missions to recommend revisions, as appropriate and necessary, to existing SSA and STM policies and regulations.

Sec. 7. General Provisions

(a) Nothing in this memorandum shall be construed to impair or otherwise affect:
  (i) the authority granted by law to an executive department or agency, or the head thereof; or
  (ii) the functions of the Director of the Office of Management and Budget relating to budgetary, administrative, or legislative proposals.

(b) This memorandum shall be implemented consistent with applicable law and subject to the availability of appropriations.

(c) This memorandum is not intended to, and does not, create any right or benefit, substantive or procedural, enforceable at law or in equity by any party against the United States, its departments, agencies, or entities, its officers, employees, or agents, or any other person.

(d) The Secretary of Commerce is authorized and directed to publish this memorandum in the Federal Register.

https://www.whitehouse.gov/presidential-actions/space-policy-directive-3-national-space-traffic-management-policy/

Index

J. N. Pelton, *Space 2.0*, Astronautical Engineering,
https://doi.org/10.1007/978-3-030-15281-9

Printed in the USA
CPSIA information can be obtained
at www.ICGtesting.com
LVHW011552160324
774658LV00003B/25

9 783030 152802